EARTH ROADS

Second Edition

About Cranfield

Cranfield University stands as one of Europe's most specialised advanced teaching and applied research centres in the areas of engineering technology and management. The University itself is unique in that most of its courses are run for postgraduates, and subsequently represents one of the largest centres for applied research in Western Europe.

Silsoe College, a faculty of the University, is a leading international centre for the application of engineering and management to the agricultural food and forestry sectors.

Earth Roads

A practical manual for the provision of access for agricultural and forestry projects in developing countries

Second Edition

John M. Morris, MBE

Avebury

Aldershot ● Brookfield USA ● Hong Kong ● Singapore ● Sydney

This book is dedicated to:

Leighton Jenkins, who drummed into my head at forestry college that most basic tenet of commercial planting: whatever you plant is only as valuable as it is accessible.

Maureen Lee who urged me to write this book and to pass on the benefit of my experience.

Mei-Ling, my wife, who had to put up with my neglectful and sometimes irritable presence, which she did with a good grace, while I retired to my study for many weeks, researched, wrote and rewrote until the job was done.

John M. Morris MBE
Sussex, England
1995

Published by
Avebury
Ashgate Publishing Limited
Gower House
Croft Road
Aldershot
Hants GU11 3HR
England

Ashgate Publishing Company
Old Post Road
Brookfield
Vermont 05036
USA

British Library Cataloguing in Publication Data
Morris, John M.
   Earth Roads: Practical Manual for the
   Provision of Access for Agricultural and
   Forestry Projects in Developing
   Countries. – 2 Rev. ed
   I. Title
   625.74

ISBN 1 85628 989 3

Library of Congress Catalog Card Number: 95-79557

Printed in Great Britain at the University Press, Cambridge

Contents

Contents

5   Maintenance

List of diagrams and tables

DIAGRAMS

TABLES

Picture groups

Foreword

However rich the potential returns from commercial plantings and other agricultural projects may appear, their value can be severely reduced if good access is not planned from the outset. This is the theme of this most welcome publication which provides a wealth of detailed practical advice on how to establish such access through the construction of earth roads. The author has, moreover, included many related activities: the taking and use of aerial photographs; quarrying of roadmaking materials; management of construction staff; machinery required and its maintenance; building of culverts and bridges; the rehabilitation of old roads, and road maintenance. The text is generously augmented by many excellent illustrations.

John Morris, after spending five years in Britain with the Forestry Commission, joined the Commonwealth Development Corporation and worked for nearly thirty years in Malaysia, Cameroon, Zambia and Indonesia, where he was concerned at first hand with the development of plantations and agricultural projects, involving land use planning, timber extraction, surveying, forest clearing, road construction and field layout. This volume distils his lifetime's experience in these fields, and forms a very lucid and invaluable textbook for all those concerned with such work, especially those in remoter areas.

The book is, as the title suggests, a very practical manual. It was first published by Cranfield Press, the publishing arm of the Cranfield Institute of Technology (now Cranfield University). Since then, with the benefit of yet more active experience in roadwork, the author has added to the original text and this up-dated and extended edition is now published by Avebury.

I would, therefore, also like to commend Avebury for publishing this second edition and wish them, and John Morris, success with this improved version of *Earth Roads*.

T.S. Jones
Chairman, Tropical Agricultural Association UK
and lately Natural Resources Adviser to the
Commonwealth Development Corporation

Acknowledgements

The author wishes to acknowledge with gratitude the considerable help given by a number of people in the preparation of this book. In particular, Mr C.J. Mettem of the Timber Research and Development Association and Mr J.V. Dishington of the Forestry Commission. Assistance was received from members of the Lake District National Park Service and special mention must be made of Park Ranger M. Guyatt who has effected a nationwide study of stone arch bridges in the UK and who provided information on the locations of suitable bridges for the relevant photographs in this book. Useful suggestions were also received from Mr P.A. Thomas of the Forestry Commission, Mr P. Easton, retired engineer ex-Commonwealth Development Corporation, Mr D.M. Brooks of the Transport and Road Research Laboratory, and Mr Leighton Jenkins, retired forester ex-Forestry Commission, all of whom read sections of the drafts and provided ideas for the improvement of the chapters submitted to them. However, responsibility for the contents of the book lies entirely with the author and the contents do not necessarily reflect the views of those named above. Thanks are also due to the Commonwealth Forestry Association for permission to reproduce information on gradients from their Commonwealth Forestry Association Handbook as a table in the text and to the Timber Research and Development Association for information on timber strength classes.

All the photographs, other than those specifically credited to other persons or organisations, were prepared from a mixture of slides and negatives, some old and some new, taken over many years by the author.

Introduction

OBJECTIVES AND LIMITATIONS OF THIS BOOK

The purpose of this book is to provoke thought amongst the managers of estates and other rural projects upon the true functions of roads in the profit producing units for which they are responsible. The book also provides some guidelines, ideas and practical advice as to how roads can most effectively be made to meet the needs of commercial projects in the present competitive environment. It is limited to earth roads in developing countries: it is felt by the writer that sealed roads are properly the province of civil engineering contractors and the mention of developing countries is as much as anything a reference to climate as, in the cooler climates of the more developed countries, unsealed roads can be badly affected by frost lift. This is an affliction not visited upon projects in the tropics and sub-tropics and so not one that will be covered in this book. That still leaves plenty of practical problems to be solved.

The book is not written by or for an engineer. It is written for practical managers who have a forestry or agricultural background and a project to create, rehabilitate or run without the aid of a qualified engineer to take on all their infrastructural problems. It is therefore pitched, to use an analogy, at the blacksmith's rather than at the mechanical engineer's level. Mathematical formulae, where they appear, have been kept simple and to an absolute minimum. The intention is to encourage the acquisition of a feel for the materials of which, and the tools with which a road, bridge or a culvert is built, and for the needs of the vehicles that use it. No less essential is a feel for the weather and the way in which rainy or dry spells can be turned to advantage or their damaging effects ameliorated.

Some degree of technical knowledge on the part of the reader is assumed. This is expected to include the agricultural techniques of the crops being grown, a basic knowledge of survey, soils, drainage and similar subjects. This book therefore does not aim to cover aspects of these related disciplines. Neither are soil stabilisation nor the use of geotextiles covered except to refer to works by other authors on the subject under the *Further reading and useful addresses* section on page 295.

THE PURPOSE OF ROADS IN A RURAL PROJECT

No manager of a railway would spend a great deal of time and money purchasing locomotives and rolling stock and then expect them to run on wooden rails. Yet, in many cases, project managers are doing something similar: they will buy their vehicles without really considering the need to match the roads to their vehicles in the way that a railwayman will suit his permanent way to his locomotives and rolling stock. It is essential to bear in mind the simple truth that vehicles, and the roads they run on, are complementary parts of one integral system and that to neglect to match them is to ensure an inefficient operation. The purpose of the road system is to be the passive part of the most cost effective integrated transport system that the management can devise.

In planning that system the management needs to weigh up the costs of the construction and maintenance of the road system against the costs of the purchase and operation of the complement of vehicles that are to run upon it, in order that the money saved on the one is not lost on the other. The aim should be to invest in a complete and reliable system that satisfies the requirements of the operation at the lowest possible cost. The result should be a compromise between creating highways, carrying perhaps only a dozen vehicles a day and accepting cart tracks over which all the project's transport has to travel as best it can. Most projects tend to spend on the vehicles and neglect the roads. It is always the easy answer to skimp on roads when the accountants put on the pressure! To them the repair of a broken down vehicle has a justifiable urgency about it that persuades them to release the cash, while repairs to the road full of pot-holes that caused the breakdown can always be delayed just so long as one can get through somehow for the moment....

The author hopes that this book will persuade some, at least, to seek and perhaps find a middle path, justified by logic.

Chapter 1

Layout

THE LONG TERM VIEW

Long established roads are only changed in their alignment after much soul searching. This is understandable, because changing an alignment can be very expensive. There will be the direct expenses in terms of the cash needed to do the job. There will be indirect expenses because of losses of planted land to be put under the new alignment, because the rehabilitation of an old road formation for agricultural use is seldom entirely satisfactory, and because of the disruption to communications likely to be suffered whilst the work is in progress. Most people therefore put up with a less than truly satisfactory route year in and year out, decade in and decade out. A search made with a critical eye can, on most projects, reveal a stretch of badly aligned road (probably unnecessarily long and tortuous) which could be the subject of improvement by realignment. This particularly applies to main roads. It is suggested that such a sector of main road is a suitable candidate for a costing exercise upon the desirability of effecting a change for the better. The costings should show on the one hand the estimated costs of effecting the change, both direct and indirect, and on the other hand the expected annual savings once the job has been done. These latter will fall into:

- Savings in road expenditures such as reduced maintenance costs, smaller area of land under roads, etc.
- Savings in road usage costs: most obviously, reduced running costs in terms of the shorter distance run by all the vehicles using it. Less obviously, savings on estates in the cumulative time wasted using it by staff and labour in getting from place to place along that particular piece of road. There is possibly even the effect of reducing the risk of accidents – and the resultant costs – by having a better and safer alignment.
- Savings in capital equipment required: realigning heavily used roads can considerably reduce the time taken to traverse them and so reduce the number of vehicles needed to effect transportation (as an example consider lorries hauling fruit along an unnecessarily long one-and-only route to a palm oil mill where large quantities of fruit are being received all day long).

1

Often the effects of carrying out such an exercise on a busy main project road are quite startling and will show both that a great deal of money has been wasted by people blindly following it, and that it does not take long to recover the outlay on a well designed realignment. Minor roads are not usually worth realigning unless this can be done at the same time as replanting, and also if little surfacing and compaction has been done so that the reversion of the old formation to useful agricultural land can be effected economically. Having carried out such a costing exercise it will be possible to see just how much better it would have been had the alignment been properly thought out in the first place.

Therefore, whenever road building is contemplated, the intending builder should consciously strive to take the long term view, particularly when a new estate development is involved.

FACTORS AFFECTING LAYOUT PLANNING

For the purposes of this book "layout" is defined as the routing and positioning of a road (perhaps in relation to the positions of other roads) as distinct from "design" which is the process of deciding how the road shall be shaped and built, and of what materials.

Two quite different types of road will be considered, namely main roads and harvesting roads (although in practice one road may serve both functions). Main roads are required to be able to carry large volumes of project traffic quickly from one point to another, whilst the harvesting road network is planned to enable the crop to be gathered in from the plantings as economically as possible. It is quite likely that the vehicles using either road will be different; all the project vehicles will probably use the main roads whereas only harvesting and staff vehicles will use the harvesting roads.

Some of the constraints affecting layout planning apply to both types of road while others apply more strongly to one type or the other. The factors, which will be discussed in detail subsequently, are as follows:

- obligatory points (particularly of main roads)
- road densities (of harvesting roads)
- permissible gradients
- anticipated vehicle types, loads and speeds, and
- anticipated future developments.

These will be dealt with firstly in the context of main roads and secondly of harvesting roads.

Main road layout

Obligatory points

The first two obligatory points will be those at which the road commences and where it will end. For example, the point on a government road where permission for the construction of the project main access route to start from has been given, and the site of the factory complex from which the project's produce is to be carried for sale to the outside world. Other obligatory points may be dictated by physical considerations (a col in a line of hills that has to be crossed over, a good bridge site for the crossing of a major river) or by social or business needs (the need to keep the road on the leeward side of a housing area, office or factory during the dry season to avoid dust becoming a nuisance to the inhabitants, or to serve other important installations en route). Before any work starts on the detailed survey of a road trace, these points should be reconnoitered and mapped. It is then possible to form some idea as to how one can lay out the road to link up and serve all the obligatory points.

Permissible gradients (see Tables 2 and 3, page 21)

It is most important that the inclines on a major road are kept within the bounds of reason. It is very common on projects to see roads driven up and down hills that are nothing short of downright dangerous. Many accidents result and, in the writer's experience, are blamed on the drivers involved for supposedly speeding when the person who had the responsibility for the road alignment is really to blame. Quite apart from being dangerous, steep hills are expensive to maintain for reasons that will be discussed in greater detail later in the book. Briefly, costs increase sharply at a point that is to a large extent dictated by the structure of the soil over which the road is built in relation to the rainfall pattern over the area. Most estates are sited on fairly well structured soils to comply with the needs of their crops and this is all to the good as well structured soils will absorb far more water before saturation and surface run-off (which will cause erosion) will occur. As a rule of thumb, as gradients exceed 1 in 15, costs begin to rise sharply and at worse than 1 in 10, become totally unacceptable both from the view of upkeep and safety. However, experience on site gained by the intelligent observation of what happens to the roads during periods of heavy rain is the best way of setting the permissible gradients, and it could well be that inclines of as little as 1 in 20 would be the sensible maximum for some of the more sandy, gritty or otherwise poorly structured and therefore more readily eroded soils. It is always better to err on the side of caution if possible. Whatever permissible gradients are decided upon, their implementation will become more and more irksome the steeper the country in which the road is to be built. It is at this point that the laying out of roads at project level, where one does not have access to the sophisticated aids of a civil engineering consultant's office (and is lucky to have good maps, or reasonably good aerial photographs and a stereoscope) becomes an art. As with most arts, some degree of flair and a great deal of dogged persistence to keep at it until a thoroughly good trace has been found and marked out, is required. During the process one is constantly weighing up, with no practical means of backing one's judgement, when one should preserve a given gradient by travelling along the side of a slope and when one should cut through a hill or fill a valley to shorten or straighten a route. Always in one's mind will be the need to cut costs and yet produce a safe and effective road.

Anticipated vehicle types, loads and speeds

The sizes and numbers of vehicles, the loads they are to carry and the speeds at which they are to operate will dictate both the necessary width of road and the minimum turning radius acceptable. To some extent the length of the project road is also a factor. If, for example, a lorry carrying produce has to cover only a kilometre or so to get onto the government main road upon which it then travels many tens of kilometres to its normal destination, it is acceptable that the project road should only permit the driver to travel at low speeds as the time lost is not very great. If, at the other end of the scale, the lorry is travelling all day on the project roads, its productivity will be seriously curtailed at low speeds. One should therefore decide whether a better road – to suit lorries travelling at up to perhaps 80 kph – or a road built to lower standards – for tractors and trailers travelling at up to 30 kph – would be the most cost effective answer.

The number of vehicles required to effect the project's haulage will give some indication as to whether single-lane roads with one-way systems or passing places are adequate, or whether proper two-lane single-carriageway roads are necessary. Note that in terms of simple earthmoving, a doubling in width of a side cut will result in a quadrupling of the amount of soil to be moved, whereas on level ground a doubling in road width means that the work is roughly doubled (see Diagram 1 below). To counterbalance that somewhat, note that the bigger the machine effecting the work the lower the cost per cubic metre shifted, and the wider the road the bigger the machine that can be used effectively.

**DIAGRAM 1   The effect on earthmoving of doubling
roadwidth on level ground and on a hillside**

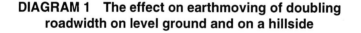
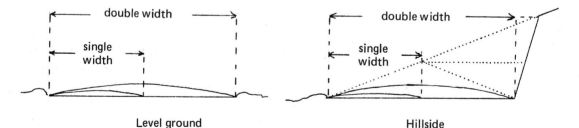

Level ground   Hillside

Anticipated future developments

The objective of this paragraph is to look beyond the development of the road system that is already formally planned for and to consider what else may reasonably be expected to occur in the future. If one is developing a tract of virgin country then it is very likely that others will develop other tracts of land beyond the boundaries of one's own estate. They may need to gain access to their land through your project, perhaps over the project's roads and, in this, they may have the backing of government or will in the future be able to get it. Indeed the government may take over one's roads compulsorily in due course. It pays therefore to consider not only what possible, as yet unplanned for developments, might occur within one's own boundaries but also what may go on outside. Those who have lived through the last few decades in tropical countries will have seen vast areas of new land opened up and the process is unlikely to slow down in the future until there is no developable new land left. There have been many cases of earlier developments blocking access to later projects whose intransigent managements have in the end had to yield to governments intent on progress, with much embitterment of relations between the parties affected, none of which does anyone involved any good. Often managers will resist outsiders using their roads because it will bring all the problems associated with heavy traffic right into their compounds which often straddle their main roads. Had they realised that this would happen and accepted it, undoubtedly they would have routed their main roads differently to leave their compounds safely on one side. When governments take over roads they usually establish road reserves over the land in which the roads run and these reserves are as a rule very much wider than the road itself so that the governments can, in the future, widen and improve the roads without hindrance from the landowners on either side. If during the process of aligning the road the estate has allowed for these eventualities the process of having one's roads taken over by the government can be fairly painless. But do attempt, when the negotiations are in train, to get a commitment from government that the road will be maintained to a standard acceptable to the project and that, in the last resort, the project has the right to upkeep the road to those standards without any hindrance from any government department. This can be very important because in times of plenty governments will probably do a good job of maintenance but in times of depression all their good intentions may evaporate, with serious consequences for the estate or project affected.

Harvesting road layout

Obligatory points

Unlike main roads, harvesting roads may have only one obligatory point, that at which the network has access to the main road. However, it is likely that there will be more than one access point to the network and that there will be some other obligatory points dictated by the need to cross physical features such as small streams and ridges or other obstacles. Generally though, one tends to be placing a network to the best advantage over a given block of plantable land, land which by its very nature, having been selected for cultivation, is not likely to provide many obstacles to road alignment. Thus for harvesting roads, the obligatory points are not likely to be so important and precisely identifiable as on main road alignments.

Road densities

Road density is defined as the ratio of road length to the area of ground served by the road network, i.e. metres per hectare (or chains per acre). The density should be given in relation to the maximum acceptable distance between roads to enable some assessment of the efficiency of a layout to be made. However, to create a truly effective total system one has to look beyond the efficiency of the vehicle and road system and consider too the efficiency of the crop extraction process from the tree or plant to the harvesting roadside. Bear in mind that the apparently simple task of removing a fruit or a cupful of latex from a tree and and putting it on the vehicle that will take it from the field to the factory is one of the biggest single direct costs of the whole production process.

Let us look for a moment at the very different characteristics of harvesting oil palm fruit and rubber latex. To keep the figures simple let us compare the following two examples:

- The harvesting of palm fruit at a yield of 30 tonnes per hectare over thirty harvesting rounds per year, which means that the palm fruit harvester has to carry out one tonne per hectare per round, almost all of it in big prickly lumps of several kilogrammes weight apiece.
- The harvesting of rubber latex and cup lump at a yield of 3 tonnes per hectare over one hundred and fifty harvesting rounds per year, which means that the tapper has to carry out 20 kilogrammes per hectare per round in the form of an infinitely divisible liquid or lumps of a few grammes apiece.

The extraction of the big and awkward palm fruit is far more onerous than the business of carrying out the rubber (neglect for this purpose the work of tapping). The implications are therefore that the maximum acceptable carry for the palm fruit harvester should be less than that for the rubber tapper on a given type of topography, also that it is more feasible to extract rubber from steep terrain than palm fruit. One can examine the tasks of the hemp, tea, coffee and other harvesters in a similar way and from that infer that the greater the quantity of crude crop to be removed per hectare per round, and the heavier and more awkward to carry are the pieces that constitute it, the more it is necessary to cut the distance over which the harvester has to carry the crop out to the roadside.

It is necessary to balance against this the value that can be placed on the loss of productive ground occupied by providing a greater road density, and the cost of constructing and maintaining the extra roads required to reduce the harvester's carry. It is not the intention here to produce a formula for calculating what is an economic carry for any given crop because, with the changing values for crop market prices and yields, land values and wage rates, any formula will provide answers that will change continually during the life of the crop. Some estates will have set their own criteria as best they can with one eye on the likely cost and difficulty in years to come of getting labour to do the more physically demanding harvesting work. Common sense is as likely to be as good a guide as mathematics.

Returning to the principles of road layout, we can regard the area to be serviced as a series of blocks surrounded by harvesting roads, or bounded by harvesting roads and other limitations such as rivers and fences. Then clearly the planting rows down which the harvester has to carry his loads should be at right angles to the harvesting roads to gain maximum efficiency. This is not always practical but as most blocks are longer than they are wide it is possible to put the planting rows at right angles to the axis of the block and thus reduce the harvester's carry to a practical minimum (see Diagram 2 below). This is of far more practical value than, for

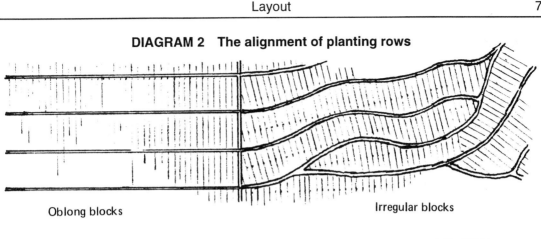

DIAGRAM 2   The alignment of planting rows

Oblong blocks

Irregular blocks

example, the long pursued habit of planting rows being aligned north-south in the hope of gaining some slight advantage in the distribution of sunlight over the trees.

On flat land with no obstacles it is possible to lay out the harvesting network in a rectangular pattern with straight roads running at right angles to one another. It is common for planters to lay out their blocks between roads in squares but this is not the most efficient method. Having decided on the maximum acceptable carry, one can double this figure to obtain the maximum acceptable distance between roads in the network. Then, theoretically, one gains the lowest density possible by having nothing more than parallel roads running across the land at this distance apart (connecting roads at right angles do nothing to improve the harvester's lot). This is of course impractical and the interlinking crossroads are necessary but should be reduced to a reasonable minimum. Table 1 below shows the effect of different length to width ratios on density in terms of metres per hectare at a maximum acceptable distance between roads of 200 metres:

TABLE 1   The effect of length to width ratios on road density

Block length (metres)	Road density (metres/hectare)
200	100.00
400	75.00
600	66.67
800	62.50
1,000	60.00
1,200	58.33
1,400	57.14
Infinity	50.00

Clearly there are substantial savings in road density to be made, with no practical reduction in accessibility for the person working in the fields, by having blocks more than three times as long as they are broad. Further savings fall off into insignificance as one gets beyond six times as long as broad. One can never reach the theoretical figure of 50 metres per hectare but with substantial blocks of flat land (several hundred hectares or more) it should be possible to return figures of 60 to 65 metres per hectare at the 200 metre spacing in a practical situation. Once hilly country with slopes exceeding the permissible gradient set is encountered, the regular blocks possible on the flat land exampled above are no longer practical. Roads will then be deflected by the need to keep within the permissible gradient limit. One can try to keep some sort of order but it is inevitable that the ideal 90 degree angle between the parallel roads and the interlinking roads will be lost and the steeper and more broken the country the more chaotic the system will tend to become. On rolling hills the density required to maintain a given maximum distance between roads will rise by fifty per cent or more and on steep broken terrain can easily exceed double the density required to maintain the standard on the flat land.

Two comments:

- Firstly, the retention of the maximum distance between roads will result in a lower average distance between roads on the hills: this provides some measure of compensation to the harvester which will be directly proportional to the extra difficulty to be encountered in harvesting in hilly country. The limit should be retained therefore to prevent the hilly areas becoming unpopular with the harvesters and the estate perhaps losing crop as a result.
- Secondly, the cost implications of high road densities in hilly places, where construction costs more anyway, will indicate very rapidly the point at which land becomes too steep for the economic harvesting of some crops. Bear in mind that as time goes on the tendency must inevitably and sensibly be to try to reduce the "hard labour" aspect of a smallholder's or estate employee's work.

Anticipated vehicle types, loads and speeds

Harvesting roads normally carry the project's slower or lighter vehicles. Ignoring staff vehicles, which have little impact, either tractors and trailers or small lorries are most likely to be used. Where large numbers are involved it is usually for short periods, temporarily, in one place (e.g. for cocoa or oil palm harvesting rounds) and the vehicles can be supervised effectively so that one-way rules can be imposed *pro tem*. Under such circumstances roads no more than twice the width of the largest vehicles (so that a broken-down vehicle does not block the movement of others) are normally adequate. There should be no compromise on the rigid compliance with the maximum permissible gradient limit. Tractors and trailers usually only have brakes functioning on the tractor's driving wheels. Whether braking or pulling uphill with full loads, their wheels soon lose their grip and, at the least, cut up the road surface resulting in rutting and subsequent erosion. At worst they may descend out of control to have accidents (which will usually be put down to "speeding" by the driver!). If lorries are to be used it should be remembered that they have wider turning circles than tractors and trailers and adequate width should be allowed for this both on junctions and sharp corners. With lorries it will also pay to make roads of greater width and to a better standard than for tractor harvesting roads so that the lorry can use the advantage of its greater speed on its way from the field to the main road system.

Anticipated future developments

One has only to consider the rotational periods of the crops planted on rural projects (timber, cocoa, oil palm, rubber and tea for example) to realise that even harvesting roads will, like main roads, be in use for decades so they too must be planned with the long-term view in mind. It is wise both to consider the approved planting programmes and to look ahead to probable developments in the next twenty years or so when laying out a network. Some of the roads that will serve for the next ten or fifteen years as harvesting roads may become main roads in twenty years time. This is not to suggest that one is immediately going to build a particular road to the specifications that may be required twenty years ahead, but to stress that, as far as can be foreseen, harvesting roads should be laid out in such a way that they can be upgraded to meet future requirements with the minimum of disruption and cost.

As with main roads, harvesting roads also may become the means of access for developments outside the estate. This may indicate the need to design in restricted access from these potential public roads to the rest of the harvesting road system in order to counter the eventual risk of large scale crop theft. If the local government has a bad reputation over the upkeep of public roads, or if the vehicle licensing or insurance costs can be reduced thereby, it may be possible, and very desirable, to lay out the harvesting system so that it can be modified to give the estate its own separate, private road system running in parallel with roads surrendered to government. This would safeguard the efficiency of the estate's transport system when the original roads have been taken over. Unnecessarily cynical? Not the least in the writer's experience. However, these are matters better judged after some discreet sounding out of the views of other developers in the district and of the relevant government departments. A little care, forethought and imagination can both save money and prevent irreparably soured relationships later.

SURVEYING IN ROAD SYSTEMS

Preliminary work with maps and/or air photos

If at all possible, get hold of government survey department contour maps, both small and large scales, and aerial photographs of the area to be roaded. These two aids each have their own particular advantages but from either or both it will be possible to pick out the main topographical features of the area under study. If the estate boundaries and the proposed sitings of the principal facilities, plantings and unusable lands can be marked on them, the approximate placings of proposed main road routes and outlines of areas to be served by harvesting road networks can be roughly sketched in. Even at this stage, some thought on the siting of obligatory points will be involved and indeed some, like the point at which the estate main access road commences and the site of the main factory complex, can perhaps be fixed immediately. For others, like river crossing points and suitably placed cols on ridges, it will be possible to narrow down the options, begin to concentrate the areas of search, and gain some awareness of the problems to be encountered in the linking together of these points by acceptable road traces.

Maps in most tropical countries will have been made from aerial photographs and the contours put in with the aid of a stereoscope. Be wary of the accuracy of such maps in countries where heavy cloud cover is common; the writer has on several occasions come across maps in which small patches of beautifully drawn contours were found to be quite fictitious: the patch presumably having been obscured by cloud on the photograph, the photo-interpreter drew in what he thought would be there and the maps were duly printed and never checked or corrected. Try matching up the edges of adjacent sheets of contour maps to see whether the contours on one sheet run perfectly into the contours of the other; if they do not, be suspicious. Even under perfect conditions the job of photo-interpretation is difficult enough. It is made very much more so in the presence of primary forest, tall planted crops or secondary forest when the aerial photographs were taken. Such cover will have prevented the photo-interpreter from getting a clear view of the ground and so the contours are likely to be in error by anything up to the height of the vegetation canopy above the ground surface. This limitation in the accuracy of the topography means that there will be no substitute for the business of getting over the ground, on one's two feet, to check everything.

In the process of sketching in possible road routes on one's map, one can note the contour interval (say 10 metres), multiply it by the permissible gradient (say 1 in 15) and obtain the minimum length of road necessary between contours (150 metres) to gain or lose height without exceeding this limit. By applying the map scale (say 1 to 10,000) one can set a pair of dividers at the appropriate measure (15 millimetres) and step off a road trace across the contours to find out whether any sort of a reasonable alignment might exist from one point to another. By this method the impossible can be eliminated and the search for a feasible trace narrowed down still further. Time spent in the office on preliminary work like this can save many days of field work for the survey gangs but should not be used as a lazy way of marking out a final trace.

Aerial photographs of just about any scale are useful provided that they are not affected by cloud to the extent that vision is seriously impaired. For estate work scales of between 1 : 20,000 to 1 : 50,000 are best as larger scales require far too many pictures to cover the ground and become somewhat tiresome to work with. The suggested scales also give a much better overall conception of land forms. The otherwise overwhelming detail visible in forest canopies, for example, tends to devolve more readily into recognisable textures which can be blocked off by texture types and sampled on site to discover what each texture type represents. Smaller scale pictures can always be enlarged if necessary and these days the definition of air photography is so good that, for estate purposes, enlargements to twice lineal size will lose no significant detail. If, for example, only 1 : 100,000 scale pictures are available, they are still worth having as a supplement to maps though they may be difficult to work with in the absence of maps.

Radar aerial imagery produces pictures from "pixells" like the pictures of a television screen. The imagery is available in both standard and high resolutions depending on the size of the pixell used – the smaller the pixell, the better the definition in the picture, and the more costly it is. Radar has the primary advantage of being unaffected by cloud. For countries where there is no cloud-free season, radar can be cheaper than aerial photography because there is no requirement for the reconnaissance plane to be standing by, perhaps for weeks, until the weather is right and normal light cameras can operate. The imagery comes in continuous strips. Vertical images give no three-dimensional effect, but side-vision image strips can be used with a stereoscope to give three-dimensional vision, as can ordinary aerial photographs. For these, the plane flies two parallel paths and these are paired subsequently under the stereoscope. Radar image definition is not as sharp as that of good aerial photographs but it is very acceptable where ordinary aerial photography is not possible.

For very large projects, in regions that have not been well mapped, investment in satellite imagery may be worthwhile. The images are not cheap but they are very, very accurate within the limitations of the pixell size used. This is rather coarse in the "Landsat" images which have been available from the United States for some years. More recently, French "Spot" imagery and Russian imagery have become available which give far finer definition. Contacts for the supply of these pictures can usually be made through the respective embassies of these countries. As an indication of the number of scenes likely to be required, Landsat pictures each cover an area 185 km by 185 km at a scale of 1 : 1,000,000. This is likely to be more than enough to contain any project in one image unless it happens to lay across the borders of two pictures. Some of the newer satellites can now give stereo vision images and show macro-features such as major fault lines and watershed boundaries that are not readily apparent at the larger scales of aerial photographs. This can be important in indicating the risks of flooding at river crossings and therefore the sizes of bridges required. Satellite images will certainly facilitate a more intelligent approach to boundary placement than the usual method of drawing straight lines across blank maps. The images can be enlarged from their negatives. However, with its coarse pixell size, Landsat imagery will lose detail at even quite small magnifications. The Russian and French, and most recently available American high-definition imagery, is more satisfactory in this respect but inevitably, substantial enlargement results in something more akin to a grainy television picture than a good photographic enlargement.

To get the full value from aerial photographs a stereoscope is essential. Stereoscopes range from small pocket units which are cheap and suitable for use in the field to very sophisticated instruments that will enable the plotting of contours and mapping directly from the air photographs to be effected. For estate purposes a simple desk stereoscope, to which a parallel guidance system can be added if the amount of work warrants it, is the basic and most essential unit. This can be equipped with a pair of binoculars for local magnification and complemented by a pocket stereoscope for field use. It is wise to get the advice of friendly survey department personnel who use these things regularly on what make and model of stereoscope to buy. It is essential to get an instrument which has the reputation of being kind on the eyes, some are nothing short of cruel to work with and lead to serious eyestrain and headaches with prolonged usage. A good unit will come complete with instructions for use and these instructions should be read and followed; skill can only come with practice.

Aerial photographs are expensive and easily damaged by careless handling. It is not generally known that they can be photocopied. While the resulting photocopies look awful, they will still give good stereo vision under a stereoscope. Much of one's rough work can be done on photocopies, either in the office or in the field where, if they are damaged or destroyed, nothing much is at stake. It is wise to record work on photocopies and retain the original pictures in a clean state as the marking of them with either crayon or ink obliterates a little of the picture and, if more photocopies are required subsequently, all the previous work will appear on them whether wanted or not. This can be a nuisance. Alternatively two sets of photographs can be obtained, one for working with, one to be kept permanently clean for unobstructed viewing. This is more expensive but in some situations well worthwhile.

Aerial photographs are usually marked with coloured soft wax or polyethylene crayons. Crayons of good quality should be used, poor quality crayons often contain impurities that can scratch the picture. Such scratches can be very confusing later when they show up after the crayon has been wiped off. Note that it is difficult to wipe matt finished pictures clean so glossy pictures are recommended. The use of ink is not recommended because of the difficulty of erasure later, though surveyors often prefer to use ink because of the fine lines that can be drawn with this medium.

With practice it will soon become possible to derive much more than just an appreciation of topography from the study of air photographs. Various textures will soon be recognised for what they are; for example even small blocks of differing clones of rubber within a multi-clone planting will become as easy to identify in an aerial photograph as on the ground. As one's expertise improves, differences in forest canopy will be detectable, not only between forest that has been affected by human interference and primary forest in its natural state, but also between primary forests with different species associations within them. These associations will reflect the local climate, the soils upon which they are growing and the types of rocks from which those soils are derived. Sedimentaries are often given away by the forest having a "planted" appearance; the trees growing best in the sandier or more friable sediments and poorest where the sediments are hard or impermeable, thus giving the appearance of planting rows. Unstable ground in the open may show crescent shaped terraces where conchoidal slippage has occurred as a result of the ground being undermined by water movement. In forest, in the absence of new slides which are immediately visible, the clue will lie in patches where all the trees lean, or have fallen, and only the younger and smaller trees are standing erect. Broad round topped ridges with narrow, deep valleys separating them will indicate well structured soils, good for making roads, whilst knife-edged ridges with evenly gradiented sides indicate easily erodible soils that will be liable to landslides, particularly from above, once a road has been cut into them. Drainage routes to enable wetlands to be firmed up, rocks or gravels in rivers and perhaps rock outcrops that have been missed by the mapmakers may well all be found on the aerial photographs with a bit of careful study and perhaps lead one to sources of road metal, saving time and money later.

It is possible that no up-to-date photographic runs are available but that there are old pictures in existence. This is no great problem if one can obtain the hire of a light plane, preferably a high wing monoplane like the Cessna 176 or one of its derivatives, and has a reasonable camera to hand. (Helicopters are not always a very good idea even if available. Some tend to wobble with a periodicity which it is very difficult for a human being to stabilise himself against and therefore pictures taken will tend to suffer from camera shake.) A thorough study of the aerial photographs first, followed by a session with the pilot and his maps as well as your photographs and stereoscope is basic and necessary pre-flight groundwork.

Arrange for the flight to take place early in the morning, as soon as the ground mists have cleared and before heat haze has begun to develop. There should be two objectives in mind: firstly to study and view the subject area with your aerial photographs on your lap to see where significant changes have occurred, and secondly to follow that up with a series of pictures to be taken of the changes that need to be recorded. Do go up high enough – most pilots seem to think that planters want to buzz the treetops. This is hopeless as what you want to see will have disappeared behind the tail and out of sight before you have any hope of evaluating it. After the exploratory part of the flight, when you start the serious business of taking your own photographs, you may need to go up to as much as 5,000 feet in order to cover enough ground with each frame. Ideally one would take one's pictures vertically through a transparent panel or hole in the floor of the plane but very few private light planes are so modified. Cessna high wing monoplanes have a great advantage in being able to fly very slowly if the pilot puts his flaps down. One can then, but ONLY WITH THE PILOT'S PERMISSION, open the side window upwards (the slipstream will hold it in that position) and get a nearly vertical picture. At least the bottom of the picture will be vertical. The result will be rather like ordinary aerial photographs that have been cut in half straight across the middle so the effect under the stereoscope will be like working with top halves only.

The pictures can be taken in two ways, either in a series with 60% or 70% overlap as is done in normal aerial photographic surveys, or in stereo pairs with 100% overlap on each pair. The first method requires more skill but is often more useful because the resulting pictures can be mosaiced; one has to hold the camera very steadily as the plane flies in a straight line alongside the sector to be covered and take a picture every time one has moved ahead one third of a picture frame. This is not easy to do well until one is used to it, so expect some erratic results in the beginning. For stereo pairs with 100% overlap, centre the picture on the primary object of interest, which should be slightly ahead of dead abeam of the plane when the first picture is taken, then take the second picture a few seconds later when the object is slightly aft of being dead abeam of the plane.

For some purposes the photographs can be taken obliquely instead of vertically and will come up perfectly well in stereo vision in the stereoscope. They are best taken as stereo pairs, not in strips. If you are photographing really low down, for example to show the height of trees, buildings or the scarp of a hill, or looking along the course of a stream or road to emphasise deviations from straightness etc., take the pictures quite close together, the stereo effect will still be good. If they are taken too far apart it will be impossible to reconcile the foregrounds under the stereoscope. Once one has become used to working with stereo pictures taken with one's own camera like this, one will soon realise that there is much more to be gained from studying one's pictures under the stereoscope than by just looking at the subject from the air and trying to memorise the results, even with the aid of a note pad. It is also much easier to show one's colleagues what one is talking about with the aid of a stereoscope and stereo pairs taken from the air which can be marked with crayon to indicate particular items of interest.

A few tips:

- Check with the pilot that the window will open fully upwards. Some newer planes have a limiting catch that prevents the window from opening until it is tight up under the wing; this will have to be removed before photographs can be taken effectively.
- Do not steady the camera on the side of the plane as it will be affected by the engine vibration and the pictures will lose their sharpness.
- Always take plenty of films, 36 exposure films if available to avoid wasting flying time changing them; take two cameras freshly loaded if you can and another person in the plane to change your films for you as you continue shooting pictures with the other camera.
- Colour reversal film is worth the extra expense over black and white as it will show up a lot more information.
- Use ASA 400 film to enable fast exposures and to reduce the blurring effect of the plane's movement.
- Make sure your exposure meter battery is fresh.
- An anti-haze filter is worth fitting in any but the clearest of conditions.
- Always loop the camera strap over your head and remove hat and glasses, if worn, BEFORE you lean out of the window to take your pictures.
- Take your surveyor and road team foreman with you if you can. They can help organise the aerial photographs, maps and films as you work and the views from the overflight will help them to understand their work later.
- When your films are developed get two sets, large postcard size or better, small pictures are not easy to work with; one set can be kept clean and the other used for marking or making into mosaics.

For the job of transferring information from one's own new photographs to the aerial survey photographs, a parallel guidance system which allows the stereoscope to move, rather than the pictures, has a lot of advantages. Such a system is cumbersome but, in the office where it should stay, this is not a great drawback. It enables one to set up two or more stereo pairs side by side or one above the other and switch rapidly from one to the other as the need arises. The job can thus go ahead fairly quickly and without too much eyestrain. New roads, new clearings and other developments can all be transferred to the regular aerial photographs with the aid of very sharp crayons and using the unchanged features in the two sets of pictures as references to locate things with. This can be a long continuing process, new developments as they are completed can be flown and the original aerial photographs updated over several years, before a new flight by professional photographers becomes necessary or becomes available from the local government survey department. Do remember that whereas the proper air survey photograph has been taken at a specific height and has therefore an approximate scale from which measurements can be taken with some confidence, pictures taken with your own hand held camera cannot be so used. Any measurements must be done on the air survey photographs or better still the maps derived from them, after they have been updated from your pictures.

Surveying equipment for estate roads

The following items are the basic requirements for a road trace survey team:

- Oil filled field prismatic compasses; not less than two, in order to avoid a team having to stop work and walk out to get a replacement if one gets broken.
- A road tracer; supplemented by an Abney level in case of emergency should the road tracer get broken but primarily to enable the team leader to scout ahead of the main party and still have some idea of gradients as he goes.
- Measuring tapes, 30 metres; far more practical to use than chains in jungle although they will deteriorate more rapidly. Always carry two per team to prevent a breakage upsetting the day's work. In many parts of the world rotans can be found and cut to length and used to save wearing out tapes. Ranging rods too can be cut from saplings in the forest.
- Geological hammer and sample bags; for taking samples from rock outcrops that might be useful for road surfacing and to assess the hardness of large outcrops. This will enable one to deduce whether they can be cut by bulldozer blades, will need drilling and blasting or whether the road trace will need realigning to avoid the outcrop altogether.
- Field notebooks, plastic bags to keep them dry in, pencils and/or ball pens, several of each.
- Marking paint; a bright mid-blue shows up best of all in high forest and secondary regrowth, bright red and white in open grassland.
- Machetes (parangs), one for each member of the party except the compassman; sharpening stones, one axe, and a chainsaw too if one is surveying through recently logged or fallen forest, in order to be able to cut through trunks and branches that would otherwise make progress very difficult.
- Adequate clothing, boots and camping equipment, a light and easily carried first aid kit.

Working with the field survey team

The team leader should be thoroughly competent in the use of the instruments listed above and able to produce maps from his chain-and-compass surveys. He should be able to fit the results of his work into frameworks derived from professionally surveyed maps of the type available from local government survey departments. He will need to be supported by two men to use the compass, road tracer and their targets and four or six men to cut the trace and the sighting rods from the jungle as they go. The gang need be no larger than this, it will be found that too big a gang is not efficient even in dense jungle because it is difficult to see far enough ahead to be able to space the workers out for safety: conversely, in light vegetation more men will not be necessary.

All the members of the survey team, the leader included, should be fit and agile: an overweight, unfit man is a menace to himself and a liability to the other members of the team. This is particularly true if the team should be out in the forest surveying road traces for days at a time and so may be many hours walk away from the nearest pick-up point. Under such conditions, an overweight individual who sustains an incapacitating injury poses a very considerable and unnecessary evacuation problem for other members of the team.

The management directly responsible for them should also be fit and able to get right into the jungle to check the team's work and provide guidance on ways and means of solving problems. The management should set the normal permissible gradient limit for the survey team to work within. It should also set the maximum permissible gradient so that, if the survey team is quite unable to find a practical trace within the limits of the normal permissible gradient, it can, without waiting for management to appear on site, make some exploratory traces at inclines between that and the maximum permissible gradient for later inspection by the management. In setting these limits it is as well to remember that during the course of construction the machines will tend to straighten out the many S-bends of the survey trace and thereby shorten it. A trace surveyed in at 1 in 15 may therefore come out at 1 in 13 by the time the road is built. This is a problem that particularly affects work done on steeply dissected terrain under dense undergrowth because one is unable to take long sightings with the road tracer. Be conservative therefore in setting the incline limits. For example, if 1 in 15 is required as the end result then, on such difficult ground, it may be as well to start at 1 in 18 for survey purposes and observe how this comes out on the finished road. Experience will soon indicate what allowance should be made during survey to get the desired end result. One can also map in all the twists and turns and then sketch in the probable alignment of the finished road and calculate the effects but this is a time consuming and tedious task. In open country where long sightings can be taken with the road tracer such problems can be alleviated by intelligent surveying.

Apart from checking that gradient limits are strictly adhered to and authorising specific steeper inclines where unavoidable, management should agree all obligatory points such as cols and river crossings to ensure that they are suitably sited given the lie of the land. One should make an effort to gain a feel for topography and for soil and rock types in hilly terrain so that one can, for example, find good firm places upon which to to site junctions, or the hairpin bends necessary to enable one to change direction on a hillside, without having to indulge in more earthwork than necessary to create a suitable formation for the purpose. This will minimise both costs and the risk of landslides occurring subsequently. Such places, which will ideally be little perched flats of some fifty metres in diameter or more, can well become obligatory points in their own right as they may be few and far between. Bear in mind that cramped and constricted junctions on hillsides can become the cause of accidents on the eventual road.

A few tips and ideas:

- In rolling country where the hills are not too high, and contour maps are probably not of great accuracy or of much help, one method of finding a trace is to start by taking a direct cutline on a compass bearing between the two points to be joined by a new road. When this is done check it with a road tracer and modify it only so far as is necessary to keep the road within the permitted incline limits. This can then be mapped out and modified repeatedly on the ground until as short and direct a trace as possible is obtained from which all unnecessary corners and unacceptable inclines have been smoothed out. This basic method can be adapted for harvesting road networks by mapping and checking deviations to ensure that distances between parallel roads remain acceptable.
- A straight line can be surveyed over a hillock that just obstructs direct vision by placing sighting rods at each end of the line and by two persons taking one sighting rod each to any two points between the end rods where each person can see, beyond the other, the rod at the furthest end from him. If each in turn then sights the other's rod in a straight line between their own rod and the one furthest away, they will both progressively move until they reach points that are on a straight line between the two end rods. Supplementary rods can then be sited within the three gaps between the original four rods as may be required. This method has its limitations but can be very useful in the field.
- The radiusing of corners can be effected by three means:
    - The first, is by simply swinging a tape around the focus of the radius at a given fixed distance and planting sighting rods, as required, to mark the corner out.
    - The second, is by taking offsets (see Diagram 3a on page 18).
    - The third, is by connecting equidistant points and placing sighting rods where the lines intersect (see Diagram 3b on page 18).

The first method is feasible with clear ground to work on and for corners of fairly small radius but the offset method is the most practical for extended corners with very wide radii.

The offset method (Diagram 3a): at the end of a straight section of road extend ahead on the same straight line the required distance between stations which we will call x (say 5 metres). Then offset, on a radius from the sighting rod at the end of the straight section, to one side, in the direction in which the corner is to turn, a distance which we will call $1/2y$ (say 25 centimetres), and put in another sighting rod. Extend beyond the offset sighting rod a further distance x in a straight line with the rod at the end of the straight section and offset again on the same side by the distance y (in this example 50 centimetres) and put in a sighting rod. Repeat this last step by aligning on the last two (offset) sighting rods, extending for the distance x and offsetting by y on each occasion until the curve is completed. It will be found that the sharpness of the corner can be increased by increasing the value of y or by decreasing the value of x and vice versa to make the curve more gentle.

DIAGRAM 3a   Radiusing corners by the offset method

DIAGRAM 3b   Radiusing corners by the intersection method

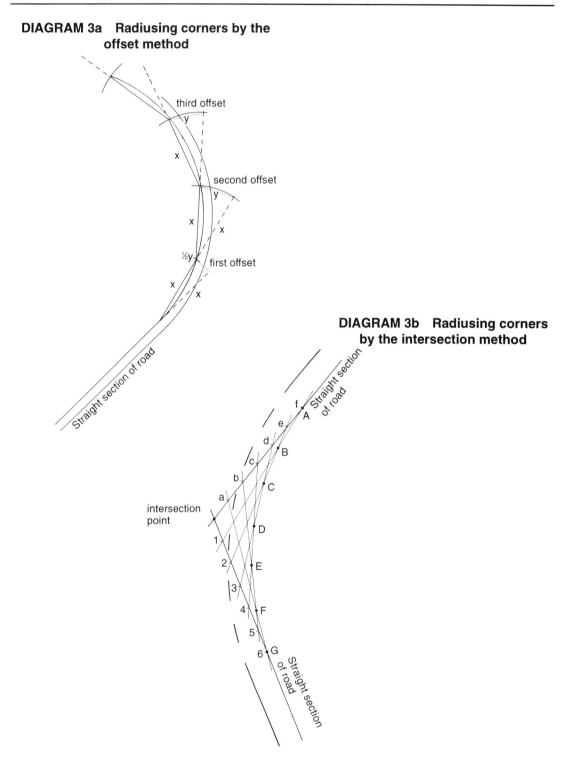

The intersection method (Diagram 3b): extend lines from the two straight sections to be joined until they intersect. From the intersection point, measure points equidistantly (say three to five metres apart) and mark them with pegs (points *1* to *6* and *a* to *f* on Diagram 3b). Then using two lines, stretch them taut between points *f* and *1*, *e* and *2*. Where they intersect place a sighting rod. Move the line connecting *f* and *1* to between *d* and *3* and where it intersects the line from *e* to *2*, place another sighting rod. Repeat this process shifting alternate lines, until the corner is fully marked (*A* to *G* on the diagram).

Always mark out the inner radius of a corner for your machines to work to, not the outer radius, because in the process of making the formation the machines will always tend naturally to throw excess earth off to the outside of the bend and to super-elevate the outside of the corner for the safety of vehicles using the road. (See also Diagram 25a on page 107).

- When selecting sites for hairpin bends bear in mind that the corner itself and the first ten metres or so leading into it and out of it should be almost level, a steep corner is very dangerous. Try to allow as generous a radius as possible, so look for sites with plenty of room. The writer has frequently seen corners both so sharp and so steep that, because of the helix effect, vehicles ascending on the inside of the bend lose all traction with one pair of diagonally opposite wheels virtually hung in mid-air, the other diagonal pair taking almost the whole weight of the vehicle. As a result ascending vehicles cling to the outer edge of the corner in order to keep going, regardless of the requirements of the highway code. This effect is purely the result of poor survey, bad design and slovenly construction and there is no excuse for it. (See also *Grading hills, corners and hairpin bends*, page 274.)

- When selecting sites for river bridges, first look at the banks of the river and their ability, or lack of it, to withstand erosion. Examine them carefully for evidence of past flood levels and damage. Look also at what is in the river bed; big rounded rocks and coarse bouldery gravel which are not native to the site will have been carried down during floods past and may indicate a potential for serious floods in the future. Next look for a little distance upstream to see whether the river is likely to erode the banks above the site under consideration as this could result in the isolation of the bridge in years to come. Finally, if possible, examine the site during a flood. Remember that almost all human activity upstream will seriously aggravate the flood risk compared with a natural forest situation. Note what quantities of driftwood are carried by the river and how big are the pieces. Consider also the effects on that situation of the development you, and perhaps others, are involved in upstream, both in the short and long term. Consider the materials available to be used in the construction of the bridge which will indicate the span length possible and whether midstream support will be needed or not. See whether piling is feasible or whether concrete foundations can be made. Where there is a massive rock bed a deep narrow site should be chosen for preference to give good flood clearance, to reduce the need for the road to descend too far to the bridge level and rise again after it, and to minimise the length of span necessary to cross the river. In areas of deeply weathered rock which has become clayey, look for a site with a straight stretch of river bed with stable banks above the bridge site. Avoid if possible, sites on alluvial soils where there is evidence of meandering and on gravelly or bouldery areas avoid braided sections of river unless there is no alternative. Always enquire from any local residents of long standing about exceptional floods in times past and their effects.

- It is important to feed back information from the field onto maps promptly and frequently to ensure that the road traces fit in with the overall plans of the estate. This is most important with harvesting road networks in hilly or rolling country as a means of controlling distances between roads. It will also enable one to respond quickly to the discovery of obstacles that may pre-empt the execution of original plans and indicate the need for revisions to keep the road density at or near the optimum.

- Before the trace has been finalised, take the man in charge of the road construction team over it and check it with him. He will probably be blessed with a lot of experience and may be able to suggest useful improvements. Even if he cannot he should be made familiar with all the obvious problems on the route such as rock faces, bridge sites, swampy areas needing filling or draining, and useful things like possible quarry sites and reclaimable timber.

- When a trace has been finally accepted it should be very clearly marked on the ground. A session on a bulldozer trying to follow a poorly marked trace through primary or secondary forest should convince anyone of the need for this. Sighting rods are not enough and trees and saplings for a metre or more either side of the road centre-line should be liberally painted (light or medium blue is recommended) to make the path abundantly clear to the operators.

- The survey team should be called during the construction of the road to check that gradients are being kept within limits. Simultaneously, if the crop to be planted will require the construction of loading platforms or landings to facilitate harvesting and extraction, the team can help with the siting of these so that they can be constructed at the same time as the road is built and in such a way as not to disrupt drainage or maintenance of the road subsequently.

New technologies

Military global-positioning satellites and their ancillary ground-based equipment are now so accurate that they could be of considerable use in surveying, in such things as road traces, planting boundaries and many other types of information required by foresters and agriculturalists. It is probable that in time all these things will become available for civilian use, at a cost that will make their use practical and economical for field operations. It will therefore pay to keep a watchful eye on developments in this field, with a view to utilising these technologies as soon as they are proven to be cost-effective.

TABLE 2   Slope and gradient

Gradient	Angle	Percent	Secant	Cosine
1 in 1.0	45.0	100.0	1.4142	0.7071
2.0	26.6	50.0	1.1180	0.8944
2.5	21.8	40.0	1.0770	0.9285
3.0	18.4	33.0	1.0541	0.9487
3.3	16.7	30.0	1.0440	0.9578
4.0	14.0	25.0	1.0308	0.9701
5.0	11.3	20.0	1.0198	0.9806
6.0	9.5	16.7	1.0138	0.9864
6.7	8.5	15.0	1.0112	0.9889
7.0	8.1	14.3	1.0099	0.9901
8.0	7.1	12.5	1.0078	0.9923
9.0	6.3	11.1	1.0062	0.9939
10.0	5.7	10.0	1.0050	0.9950
12.0	4.8	8.3	1.0035	0.9965
15.0	3.8	6.7	1.0022	0.9978
20.0	2.9	5.0	1.0012	0.9988
25.0	2.3	4.0	1.0008	0.9992
30.0	2.0	3.3	1.0006	0.9994
33.0	1.6	3.0	1.0004	0.9996
50.0	1.2	2.0	1.0002	0.9998

Note:
- Angle shown in degrees and approximate decimal above horizontal.
- Secant gives slope distance per unit horizontal.
- Cosine gives horizontal distance per unit slope.

(Information above derived from the Commonwealth Forestry Handbook of 1981 and presented by courtesy of the Commonwealth Forestry Association.)

TABLE 3   Gradients in degrees and minutes

Gradient	Angle	Gradient	Angle	Gradient	Angle
1 in 1 =	45° 0′	1 in 11 =	5°12′	1 in 25 =	2°20′
1 in 2 =	26°34′	1 in 12 =	4°46′	1 in 30 =	1°54′
1 in 3 =	18°26′	1 in 13 =	4°24′	1 in 35 =	1°38′
1 in 4 =	14° 2′	1 in 14 =	4° 5′	1 in 40 =	1°26′
1 in 5 =	11°18′	1 in 15 =	3°48′	1 in 50 =	1° 9′
1 in 6 =	9°28′	1 in 16 =	3°35′	1 in 60 =	0°57′
1 in 7 =	8° 8′	1 in 17 =	3°22′	1 in 70 =	0°49′
1 in 8 =	7° 7′	1 in 18 =	3°11′	1 in 80 =	0°43′
1 in 9 =	6°20′	1 in 19 =	3° 1′	1 in 90 =	0°38′
1 in 10 =	5°42′	1 in 20 =	2°52′	1 in 100 =	0°34′

Picture Group 1:   Surveying

1.1   Good reason to make a clearly marked trace: when pushing through forest like this it is very difficult for the bulldozer operator to see an ill-marked cutline and follow it.

1.2   Good reason to keep gradients down to sensible inclines: every wheel mark is a potential cause of erosion when the hill is too steep. This makes upkeep very expensive to say nothing of the nuisance of getting stuck on it if it rains.

1.3 A trace cut in secondary bush regrowth is fairly easy for a bulldozer operator to follow though it can be the most difficult vegetation for the survey party to work in.

1.4 Secondary or primary forest is much easier work for the survey team but difficult for the bulldozer operator to see his way through. The one colour that really stands out is sky blue, perhaps because it is the one colour that is rarely found in the forest naturally.

1.5   An Abney level can be the basis of a simple road tracer to maintain an even gradient. A staff with square cut ends is used to hold the Abney level on and another staff, slightly taller, is cut and the top split to receive a target, the centre of which is level with the centre of the Abney level sight tube.

1.6   The target consists of a small piece of cleft wood and its correct position in the target staff can be marked with a couple of pencil lines in case it gets knocked out of place.

1.7a/b   The Abney level is marked in degrees above or below the horizontal whereas the Orr's Ghat and road tracer in use in the lower picture is marked in gradients (i.e. 1 in 15). The Orr's Ghat and road tracer is more accurate than the Abney at the gentler gradients but the Abney is much easier to use in forest because it is less cumbersome and because a broken staff is easily replaced. With care in setting the Abney, good results at inclines between 1 in 20 (2°52′) and 1 in 10 (5°42′) can be attained, these being in the range of greatest interest for road building.

1.8   The working part of an Orr's Ghat and road tracer. (See Diagram 4, page 28.)

1.9a/b   Observation of the way in which soils bared to the elements react can be revealing: typically pale coloured soils which are sculpted by the rain into miniature Grand Canyon-like formations as in Picture 1.9a (top) are poor road making material and very readily eroded. Typically brown or red-brown soils that remain blocky or break down into fine crumbly surfaced rounded formations like that in Picture 1.9b have good crumb structure, will compact and bear weights well and are relatively resistant to erosion. It may be necessary to blanket the poor soils with good soils to establish a road formation in which case survey teams should be able to recognise their different qualities and point out their existence to management.

1.10 Rocks like this, core boulders, the results of colloidal weathering, are very difficult for bulldozers to remove and can cause much wasted time unless rockdrills and explosives are used. Survey teams should learn to recognise this sort of obstacle and avoid it if possible when laying out road alignments. If not avoidable, management must take account of the problems that will arise.

1.11 Volcanic ashes in the side of a long extinct volcanic cone: the pale layers and those between them and the soil have weathered and become cemented to the stage at which they were very hard to cut into with a bulldozer blade. Below that level the ashes are as clean, loose, sharp and dry as they were when they fell from the rim of the cone. They make excellent road gravel when scarified in with the heavy clay soils of the district. Similar cones are frequently found in areas where there are basalt soils and they show up clearly on aerial photographs. One such deposit can provide all the road gravel that any project is likely to need.

DIAGRAM 4   Orr's Ghat and road tracer

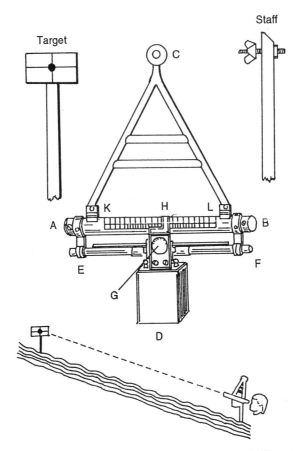

This is a useful instrument for carrying out rough preliminary survey of hill route of any required gradient.

It consists of a hollow metal sighting tube A B fitted with an eyepiece at one end and cross wires at the other. The tube pivots round the point C, which is held suspended from an upright staff. The required inclination is given to the sighting-tube by means of the weight D, which is capable of motion along the rack E F by means of the screw G, H is an index which moves along a graduated scale of gradients K L marked on the body of the tube.

Thus, supposing a gradient of 1 in 43 has to be laid out along a hill slope. The weight D is moved along until   H   reads 43 on the scale. A man with a target, which is set at the height of the axis of the sighting-tube above the foot of the suspending staff, is then sent along the hill-side to a convenient distance — usually from 50 to 60 yards, until the place is found where the cross wires intersect on the centre of the target. The foot of the target is then resting on a point, from which the slope down to the observing point is one of 1 to 43. The assistant drives a peg in here, to which the surveyor advances and repeats the operation. In the figure the target is shown as turned through a right angle.

Chapter 2

Materials

MAKING USE OF WHAT IS THERE

Soils

On rural projects the native soil itself will form the primary building material for the roads. It is important therefore to be able to recognise which soils are suitable and which soils are likely to give trouble. Fortunately most soils are good for agriculture for the very same reason that they can be good for road making: namely their structure. Well structured soils have excellent bearing qualities when normally moist, though they will break down under heavy loads if either too wet or too dry. Best are the heavy, well crumb-structured tropical clays derived either by weathering in situ, or by weathering, erosion and redeposition elsewhere, from ultra-basic and basic volcanic rocks and limestones. They range in colour normally from dark chocolate brown through medium browns, pinky browns and reds to deep cerise reds. If waterlogged they may be grey or grey-blue but this will be a colour local to the waterlogged site. Where unaffected by waterlogging they will be their normal colour. Such soils are easily worked, firm up readily if not puddled and will not erode easily once compacted. Whitish and pale yellow structureless clays and black montmorilonitic clays can be very difficult, they will be crisply hard when completely dry and like toothpaste when wet – indeed sometimes it seems they have no intermediate state when they are workable! Sandy and silty soils may be quite unable to bear weight unless very wet, like sand on a seashore, and will erode very readily. If they have a white or pale yellow clay fraction in them they may be firm enough to use as a sub-grade, provided they are well blanketed by better soils to give reasonable bearing capacity but unblanketed they will erode easily, become plashy on the surface after even quite small rainfalls and will not bind a river gravel or crushed stone dressing well. Colour is not the only guide; observation as to how the soil reacts to being bared, turned over and left to the elements, mainly sun and rainfall, will give further hints (see Pictures 1.9a/b on page 26). Soils that break down into smooth crumbly surfaced miniature downland-like landscapes are those with good structure. Those that erode rapidly into miniature Grand Canyon-like landscapes are poorly structured and are not the best of materials for estate road making. These are two extremes and there will be variations across the range. Taking handfuls of these soils under normal conditions of moisture and compacting them, breaking them down and repeating the process will soon give a feel for the good and bad soils which will, in the long run, be the best guide of all.

Estate managers seldom think it is worthwhile to move and use good soils to blanket-cover sectors of roads on poor soils but it is an effective answer to the problem (given that drainage has also been properly done) and is cheaper than using great thicknesses of gravel or stone. The use of coarse stone is also overdone on many projects and can lead to a lot of costly trouble later for the maintenance grader. Good soil on the other hand is never likely to damage a grader and will hold a surface dressing of gravel well, making a road that is a pleasure to drive along and easy to maintain. When road construction is in progress it pays to note where cuttings in hills expose and can yield blanketing material for use over poor soil or to top off fills over swampy sectors. In effect, use the cuttings as temporary quarries from which to draw good soil.

Sources of gravel and stone

It is in the writer's opinion a great pity that some knowledge of the dynamic geological processes that have formed, and are still forming our landscapes, is not included in the average school curriculum, and is often only very inadequately covered in agricultural courses at degree or diploma level. A knowledge of these basic principles is of immense help in knowing where to look for gravels and stones for road making. It is not the brief of this book to remedy this shortcoming but to suggest that managers who wish to be proficient make the effort to learn something of the subject in their spare time and remain particularly observant of rocks and topography during their work-time. It is also regrettable that when drawing up the boundaries of agricultural projects, the needs for quarries are often completely overlooked and, as a result, suitable outcrops are deliberately excluded because they are obviously not plantable. The resulting project then finds itself in the position whereby it has little or no stone of its own and has to buy it from a few yards outside its boundaries at an exorbitant price from someone who has had the sense to realise that a strategically placed bit of "poor" ground is worth a fortune.

Even under primary forest, volcanic, sedimentary and alluvial landscapes of various types show up very clearly under the stereoscope when studying aerial photographs, and these can give indications of where to look for useful deposits. Volcanic cones are made of ashes – these may have become very weathered and hardened on the outside but after penetrating in for a few metres, loose fresh ashes may be found. These are a superb road surfacing material, sharp, dry and fine, giving a good grip, mixing well with clay soils to form a durable but easily worked and graded surface. Look also for the talus or fan deposits of mountain streams where they disgorge onto flatter areas of land. They probably consist of still unweathered river gravels and stones brought down during the floods of centuries past. They will not perhaps be well graded for size but, if not compacted or mixed with too much clay, may be cheaply screened to make them more readily acceptable. River beds may provide gravel that can be dredged for the purpose but gravel dredging activities may upset people downstream who draw water from the river and special permission may have to be obtained to avoid trouble arising. Seashores may be a source of shingle but most shore lines are protected and even if they are not, the removal of gravel in quantity may destabilise a coastline very easily so it is not a source to use, except where deposition clearly exceeds erosion, or in an emergency.

Possible quarry sites for gravels formed from raised beaches, weathered conglomerates or rocks that are soft enough to be ripped by bulldozers may have been found during road surveys. More may be found during the actual process of road building when the bulldozers are cutting into hillsides. Opportunities like this should be exploited when they occur and can be of great value to the lucky estate possessing them. Growers and processors of oil palm fruit will have their own source of "gravel" in the form of excess shell discarded by the factory and a very good surfacing material it makes too!

Crushed stone

The use of crushed stone is not a cheap option. Unless a large deposit of rocks of a size fit to pass through the jaws of a crusher are available, drilling and blasting will be necessary. To do this an adequate air compressor and drills will be required as will the use of explosives. Be warned: do not be tempted to save money by getting a small compressor which has apparently just the right capacity, it will in practice be found wanting. In most countries quarrying with explosives will necessitate the employment of qualified operators with licences and the provision of proper explosives stores and permanent security guards. It is likely therefore that most projects will be only too pleased to hand this part of the operation over to a contractor with the necessary skills who is acceptable to the local police and security authorities.

There are small mobile crushers of about 4 tonnes per hour capacity and suitable for hand feeding on the market but, if the project has a substantial programme, it will be necessary to consider the purchase of substantial machinery. This may be mounted on a chassis on two or more axles or only mobile to the extent that the engine and crusher are mounted on a skid-shaped base with supplementary machinery that can be dismantled for transfer between sites and then reassembled. However, crushing plant, big or small, is expensive. This could mean that investing in the plant necessary to produce crushed stone will only be viable where either no other source of stone is available at less cost, where crushed stone is required for the construction of some other major works on the project – the combined requirement justifying the capital outlay – or where access exists to a profitable market for crushed stone that would make the investment worthwhile.

Apart from the direct cost of making the crushed stone there are other costs:

● Crushed stone can be very sharp when newly applied to a road and can be cruel on the tyres of vehicles running on it thus increasing vehicle running costs.
● Crushed stone will require proper mixing with the soil on the road to bind it into a durable surface whereas most gravels, excluding river gravels, will contain their own binders.
● Because it is likely that the project will have only one crusher, average delivery costs to the point on the road system where the stone is required will be greater than would be the case if it were possible to make direct deliveries from the nearest of several gravel quarries scattered around the estate.

Because of these points it may be better to apply coarse but easily broken stone to the roads, if it is available, and crush it in situ with bulldozer tracks, a vibratory sheep's foot or a grid roller (compactor) and accept the trouble that this makes for the grader during road maintenance. Even allowing for the extra cost of effecting the compacting required this may well be the cheaper option in the long term.

Building stone

Hard sandstones and other sedimentary rocks showing a pronounced layering along which they can be cleaved can make very useful stone for the construction of bridge abutments and piers, stone arch bridges, revetment walls and other structures.

Bridging and culverting materials

Where forest land is being converted into plantations it is possible that the heavier and more durable timbers, which may be difficult for the loggers to market, will be left behind and can be used for bridge and culvert construction. Many species will last for decades, some for more than a century. Management should therefore be knowledgeable of local timbers so that they can recognise and utilise those that are suitable. Note that some timbers that are susceptible to rot when alive are quite durable when dead and buried. Hollow logs of these species can be cut to length, then if the hole is not clear enough, clean it as far as possible with poles to the point at which it can be seen through. Place the log at an incline on an earth slope and light a fire with lots of dry branches in the hole. When the hole can be seen to have burnt clean, cover the bottom of the log with earth to kill the fierce draught of the fire and plug the top end with a piece of green wood and seal it with wet mud. Leave it to cook like this until the fire is well and truly extinguished, a process that will improve the durability of the wood, and then clear out the ends to get a ready made small diameter culvert pipe that will last twenty or thirty years (see Picture 5.5 on page 138).

Where drilling rods and explosives are available, the beds of rocky streams that have to be crossed can be cleared to improve the flow of the water by blowing up the boulders in the centre of the bed. The rock shards resulting are excellent material for building dry stone walls (they can be cemented if necessary) for culvert or bridge abutments.

BOUGHT-IN MATERIALS

Unless they have the equipment to log, prepare and mill their own timber, it is perhaps better that project managers should put major bridge building out to contract rather than buy in the materials and build for themselves. However, culverts should be within their capability and, for this purpose, the purchase of steel or concrete culverting is the most practical course. Regrettably some countries ban the import of nestable steel culverting to protect local cement piping industries. If the two are available, and the water to be passed through them is not particularly corrosive, the steel culverting, up to about two metres in diameter, has the following advantages:

- It is far easier to carry to site than the equivalent concrete culverting (an ordinary pick-up truck can carry in one load of steel culverting the equivalent of two or three lorry-loads of concrete culverting).
- All the pieces of steel culverting are light enough to be manhandled into position. Concrete culverting would require lifting tackle for all but the smallest sizes.
- Steel culverting is less easily damaged, both in transit and during assembly, than concrete piping which is particularly prone to damage around the seals.
- There is a degree of flexibility in a completed steel culvert which will tolerate at least some settlement without damage. There is no such flexibility in a concrete culvert.

For estate purposes, where stream flows require the provision of more capacity than is available from a two metre diameter culvert or a battery of culverts, a bridge is probably cheaper and more suitable. One of the practical advantages of a bridge – with its separate abutments or piles – is that the stream bed underneath it can often be deepened to some extent without affecting the bridge. With a large culvert this is not possible without indulging in a major operation, because the culvert's floor forms the bed of the stream. This can be an important factor if future land drainage improvements are contemplated.

It is therefore likely that purchase of culverting for estates will most commonly fall into size classes between half a metre to one and a half metres in diameter.

Wire for home-made gabion baskets, bridge beam and capsil lashings and other similar purposes, should be of soft iron, well galvanised, and can be a fairly large stock item in certain circumstances. Concrete reinforcing mesh can also be used for gabion baskets. Mesh, reinforcing rods for concrete work and netting for use for the establishment of ground cover to counter surface erosion should all be stocked. Cement and sand for concrete are better purchased fresh if possible because cement can deteriorate with storage and sand is difficult to keep clean.

Cement, culverting, reinforcing metal and wire apart, one should try to avoid having to buy in major road supply items. Self-sufficiency should be aimed for to avoid being at the mercy of suppliers of things like timber, sand, gravel and stone.

Chapter 3

Machines

TOOLS OF THE TRADE: SELECTION AND EQUIPMENT

If care in the selection of precisely the right machines for the job in hand is important for a big-budget project, it is *vital* for the small project where perhaps only two or three machines have to cover the whole range of work to be done. All earth-moving machinery is very expensive to purchase and run: an ill-considered purchase can be a major setback.

Reliability, on all projects, must be the prime consideration when making a choice, not just the inherent reliability of the machine itself but of the back-up service that goes with it. Wear rates on tracks, track assemblies, cutting edges and many other parts are high and it is essential that your supplier can provide you with parts and skilled mechanics to keep the machines running. Machines lying idle awaiting parts, and/or service, cost money in depreciation and finance charges, the costs of operators standing idle and also in the losses to be experienced when road building programmes and development programmes are adversely affected. Plantings may be delayed, cost more or even be made impossible with, perhaps, the loss of a season's nursery stock or the reversion to weeds of land cleared in anticipation. A good supplier can therefore make the difference between success and failure.

Versatility, for small projects, must be the next consideration. Where a big project can, for example, fully employ several bulldozers for earth shifting, front-end wheel loaders for loading gravel in quarries, and so on, the small project may need to compromise with, perhaps one tracked front-end loader equipped with a ripper. It will shift earth – not so well as a bulldozer – it will load gravel – not so fast as a wheel loader perhaps – but it will do both jobs competently on a small scale. Thus the choice of exactly the right machine for the job is often more critical for the small-budget operation than the big one. Avoid getting "conned" into using cheap "bolt-on" conversions for agricultural tractors – they will almost always fail when the hard work starts.

Bulldozers

These machines are available in an immense range of sizes from about 30kW (40 hp) to 600kW (800 hp), weighing from 3 tonnes up to 80 tonnes. There are two principal manufacturers, one American, the other Japanese and a large number of smaller manufacturers. Most estate requirements will be adequately covered by machines between 60kW (80 hp) and 180kW (240 hp) although a very large project may find justification for one or two of the heavy bulldozers rated at around 375kW (500 hp).

When considering the size of machine, one should tend to go for a size larger than the obvious requirement. Bulldozers have a rough life and it is far better that a machine should be working comfortably within its capability than that it should be straining to do work too heavy for it. Having said that, one also has to remember that big machines move vast quantities of earth very cheaply but are not able to get around nimbly in forest, particularly where topography is steep and there are rocky outcrops. For this the smaller machine is quicker, more stable and safer. One may therefore require a mix of sizes for the work in hand, something that is more easily justified on bigger projects.

The basic machine being so costly, it is false economy not to equip it carefully with the right options. Perhaps the first of these is to choose between direct-drive and power-shift gears. Power-shift as offered by the premier makes is now so good that it is often more reliable than direct-drive, because the operator is in no position to misuse the gear shift. However, direct-drive is still very popular amongst loggers for log extraction machines where the country being worked is hilly. For earth moving, power-shift is usually preferable. Some manufacturers are providing bulldozers with hydrostatic drive: forward, reverse and turning are all controlled by a joystick. There are no gears to change. These machines are supremely easy to operate and very fast working. They have the additional advantage of driving both tracks when turning a corner, whereas conventional machines with steering clutches lose power on one track when cornering. If they are available in the sizes required they are well worth considering. However, check the back-up service. Do try the machine out before purchasing.

The next item to look at is track width. Narrow tracks are desirable for very rocky ground to prevent the track plates from curling up at the edges under the weight of the machine. Standard tracks are acceptable where the ground to be worked is all firm but usually two sizes of broad tracks will be offered and the one median between the standard and the broadest will suit most rural project situations. It will give that extra bearing that will enable the tractor to negotiate the occasional piece of softish ground and yet not be too liable to plates curling when the occasional rocky patch is encountered. If, of course, your ground all tends to be soft or has no boulders in the soil; expansive alluvial flats for example, then go for the broader tracks. Swamp tractors are also available and these have specially constructed undercarriages to cope with the very wide tracks required to enable them to operate in marshy conditions. They are designed for this purpose and are not suitable for working on ordinary ground. Various track shoe designs are available. The standard single grouser shoe will suit most estate work. Rock shoes are really only justified in very difficult stony conditions and, unless the bulldozer is required to work most of its time in a stone quarry, these are unlikely to be suitable for an agricultural project.

Track guards are available to prevent stones from becoming trapped in the track chains and rollers or sprockets. If the machine is to work in loose gravel, particularly when working to gather gravel from a river bed for example, they are worth fitting to reduce wear and tear.

One manufacturer has introduced the use of high drive-sprockets (see Picture 2.3 on page 45). The objective of the change is to reduce the risk of hard stones or other objects being caught inbetween the tracks and the sprocket, where they can severely strain the track tensioning equipment or break the sprocket teeth, and also to generally reduce wear and tear. It is particularly worth having if the bulldozer is to be used in stony conditions, for example, moving stones and gravel in a fast-flowing river bed. It also brings the dozer blade much closer to the machine, making it easier for the operator to control and make a more even cut, which is particularly helpful when working heavy clays. If the price penalty is not too high, it is an advantage worth considering when comparing the offerings of different manufacturers.

There are various blade shapes available but of these the angle blade (A-blade) and straight blade (S-blade) are the most commonly used for general earthmoving. A-blades have three setting positions, one at right angles to the centre-line of the tractor (the straight position) and one each side at 25 to 30 degrees from the straight position. To be able to set the blade at an angle is particularly useful when cutting along the side of a hill as it enables one to throw the earth off down the hillside more easily. The changing of the angles is usually done manually. At time of writing not many manufacturers make machines equipped with hydraulic angling. This is not surprising because the angling arms are easily damaged and the cost of replacement of a hydraulic unit would be very high. Both A-blades and S-blades can be tilted to raise one blade corner and lower the other either way. For this hydraulics are commonly available and fitting them can markedly increase a tractor's output, particularly for forest clearing and road work. Manual blade tilting is very time consuming and the turnkeys provided for so doing tend to be more readily damaged than the hydraulic units, in the writer's experience. A-blades are wider but not so deep as S-blades. If the project is only going to need one or two bulldozers, then it is better to specify S-blades with hydraulic tilts, but if a larger number of bulldozers is required and the ground to be worked tends to be steep, then A-blades are worth having on some of the smaller units.

On all but the easiest of conditions heavy duty blade corners are better than the standard corners. They will help considerably with the felling of trees and the removal of tree stumps and rocks.

Where the clearance of old rubber or other tree crops constitutes a fair proportion of the work to be done, the use of a tree pusher mounted above the blade may be justified. However – a word of caution. A tree pusher has to be long enough to push a tree to the point at which its root plate has lifted from the ground before the blade reaches it, otherwise the dozer risks being snagged when the roots lift (as the tree falls) and force their way up between the blade and the radiator. Extracting a dozer from this situation can be time consuming and damage to the radiator may occur. A tree pusher that is too short is also inefficient as the dozer will need to push the tree part way with the pusher, back off, and then finish the job with the blade instead of pushing with the pusher, following through and extracting the root with the blade in one smooth movement. To be long enough to be efficient the pusher will be heavy and will shift the centre of gravity of the bulldozer far forward unless it is counterbalanced by a winch or ripper. Despite the foregoing, where old rubber has to be felled and cleared, a good tree pusher on firm ground can be very efficient indeed, especially if followed up with chainsaws to cut the branches at their joints and S-blade dozers to windrow the resultant brash. For this work, machines of around 150kW (112 hp) are required.

Most bulldozers can be fitted with either a winch or ripper. If your project has soft ground that your dozers will have to cross or work, then the first two machines should be fitted with winches to facilitate recovery of one bogged machine by the other. In firm conditions, the decision as to whether to fit winches or rippers will depend on the work to be done but even if there is no obvious need for winches, one or two machines at least should be so equipped as there will certainly be occasions when a winch will be found to be worth its weight in gold. The winches supplied by the premier manufacturers are thoroughly satisfactory, standard pieces of equipment, the only variable being the thickness of the wire ropes used. For occasional use, and where long reach is not specially required, the thicker of the two options normally available should be specified. Rippers are offered by the two big makers with either a single large shank, for very heavy ripping of rocks, or with three shanks for normal ripping. Since it is often necessary to rip stiff tropical clay soils to speed up earthmoving output, the three shank ripper will be the usual choice. For particularly hard or stony conditions the operator can remove one or two shanks as appropriate and replace them when normal work is resumed.

The fitting of a rollover cab to any bulldozer is recommended. Where the machine is to work in forest or do any felling, then side plates should be fitted to prevent damage to the engine compartment. In the tropics most bulldozers are supplied with pusher fans which direct the hot air from the radiator away from the operator but sometimes tractor fans are fitted – it is as well to check.

Bulldozers are most efficient at cutting and moving earth for relatively short distances – up to 100 metres, say. For longer distances major road builders use scrapers which may be towed by bulldozers or self-powered. Big ones can load and shift up to 20 cubic metres (26 cubic yards) of soil at a load. As they plane off a thin layer of soil over some distance during loading they are not suitable for rocky or bouldery ground. This limitation, together with their expense and lack of versatility will probably make them unsuitable for estate use, unless the project is a very large one on rock free soil. A more versatile method is to supplement a dozer with a front end loader and dumpers to effect long hauls of soil (to make an embankment over a piece of low lying ground for example). The dozer can rough cut and heap the soil for the front end loader to put into the lorries. The loader and lorries can be used for spreading gravel on prepared formations at other times.

Graders

There are a number of adaptions and towed versions for fitting to agricultural tractors for grading roads. In the writer's experience, only one has proven suitable and this unit required a 60kW (80 hp) four-wheel-drive tractor to operate it at the very minimum (see Pictures 2.16 & 2.17 on page 52). It was effective for maintenance but not really adequate for road construction. It is better therefore for a purpose-built machine to be obtained. The big manufacturers offer machines from about 48kW (65 hp) to 260kW (350 hp). As with the bulldozers, it is a very wide range. For estate use, the size of the harvesting roads will probably limit the size of grader that it is practical to use because the harvesting roads will constitute by far the greatest length of road to be built and maintained. If the project is very big it may justify the selection of a large unit for the main roads and several smaller units for the harvesting roads. Bear in mind that a small grader can finish or maintain a big road but it will be rather slow at the job. A big grader cannot maintain a small narrow road, it simply has not enough room to operate in. There is not quite the same question of the machine having to be adequate for the job in the way a dozer has to be. Indeed, skilled operators of the smaller dozers can do a very good job of finishing off a new road with well formed cambers and super-elevations which the grader really only has to smooth out so, that if there will not be long term need for a large grader on a project, it is better to select smaller machines that can efficiently operate on the harvesting roads. However, even while bearing these points in mind, it is stressed that one should choose an adequate machine in terms of power and size and work it comfortably within its limits rather than put up with something that is too light and flimsy. For big projects with substantial main road upkeep requirements, machines of about 100kW (135 hp) are suggested. Smaller projects with little main road to upkeep might cope with a 60kW (80 hp) unit but this is really about the smallest size that will be practical.

It is very important to equip the machine properly. There are a lot of advantages to having an articulated frame (or rear wheel steering) as this will enable the machine not only to operate more efficiently and give a bigger work output, but to crab its way out of the sort of trouble spot that a fixed frame machine would get bogged down on – a soft road edge for example. This can save having to send a dozer, perhaps from some considerable distance away, to rescue it. Articulated machines can turn very much sharper corners than rigid framed machines and so cope better with operating in smaller roads. It is important to check that the front wheels are capable of turning sufficiently to more than compensate for the articulation so that, for example, if the frame is articulated to the left, the machine can still follow a right-hand curve in that configuration (see Pictures 2.13 & 2.15 on pages 50 & 51).

Most of the graders made by the large manufacturers mount their blades on a turntable, which can swing right out from underneath the frame to one side or the other as required (see Picture 2.11 on page 49). This enables the machines to grade the sides of cuttings to give a smooth finish. However, estate roads will tend to be superficial for the most part and this facility is seldom, if ever, used. The turntable is probably the most vulnerable part of the machine if it is abused and is expensive to replace. Some smaller manufacturers of lighter machines do not use turntables (see Pictures 2.13/2.14 & 2.15 on pages 50 & 51 and compare with Picture 2.8 at the bottom of page 47). Turntableless machines are nevertheless effective for estate maintenance.

Most mouldboard blades have hydraulic side shift these days – but it is worth checking that it is provided. It is a very necessary facility.

To assist in cutting compacted road surfaces, most graders are equipped with a scarifier mounted forward of the mouldboard. This is not ideal as it can obstruct blade movement. To counter this effect the scarifier is usually rather small. Its biggest drawback is that it cannot reach out to the width of the grader's wheel track and therefore hard ground on the edges of the road cannot be ripped by it. This means that any hard ground at the edges has to be cut by the mouldboard blade, unassisted, at its outer extremity where it is at its weakest due to the effects of leverage. It is better instead to specify a rear mounted ripper, usually with five, seven or nine shanks, which will loosen ground right out to the edge of the grader's wheel track. This will leave the operator with no excuse for straining the mouldboard and turning circle mechanisms (see Pictures 2.7/2.8 & 3.5 on pages 47 & 73).

Some grader manufacturers offer a front mounted dozer blade which can be fitted forward of the front wheels. This has several advantages:

- The front blade can be used to clear road drains leading off the main formation. This the mouldboard blade would be unable to complete because the front wheels would drop off the edge of the formation beforehand, leaving the grader suspended on its blade.
- The front blade can be used to clear odd rocks or tree trunks that have fallen onto the road as a result of landslides without the need to call for a bulldozer (see Pictures 2.6, 2.10, 2.13 & 2.15 on pages 46, 48, 50 & 51).
- When tippers dump gravel it is frequently left in big heaps which the grader, unequipped with a front blade, has to scramble over. For a grader with the front blade set, say, half a metre above road level and the mouldboard blade set somewhat lower still, the spreading of such heaps becomes easy (see Picture 3.6 on page 73).

CAUTION: these blades, although called dozer blades, are not strong enough to emulate a bulldozer in action and operators need to be made to understand this quite clearly.

Rollover cabs are available for graders and their use is strongly recommended. Fully air-conditioned cabs are also available and where there is a lot of maintenance to be done on heavily used main roads they will protect the operator against the dust raised by passing vehicles and greatly increase his output. Under such conditions the provision of what might at first glance seem to be a luxury item will soon pay for itself. Non-believers should be invited to try operating a grader on a busy, dusty road: conversion is guaranteed!

Tyre inflators are often offered as an optional extra, they are worth having and can save the wasting of many hours over small punctures although they are of little help with big ones. If several graders of the same model are operated, the purchase of a spare wheel, kept ready for emergencies at the workshops, is recommended.

Lighter machines can be "stood up" on their hydraulics (see Picture 2.13 on page 50) which is valuable for a quick wheel change in the event of a puncture, but DO NOT ALLOW ANYONE TO GET UNDER THE MACHINE, AND ALWAYS BLOCK OR JACK THE AFFECTED HUB BEFORE REMOVING THE WHEEL.

Compactors

The two primary functions of compactors are the compaction of earth fills and the compaction of road surfaces and surface dressings. Occasionally the compaction of the beds of pipe culverts is also practical. First choice will almost always be for a smooth steel roller and where only one unit is required this is the best choice. The question then is whether it should be towed by a tractor or self-powered, and if self-powered, of what form. Towed rollers are of course cheap but they will tie up a tractor when they are in use and, if ballasted with water to make them really effective, are surprisingly difficult for a wheeled tractor to handle in anything but the easiest of circumstances. It is often necessary to use a crawler to haul them effectively, particularly for construction work. Being towed, the total operating unit is long (tractor plus roller) and cumbersome to use, particularly where room to manoeuvre is restricted. The tractor driver needs to be skilled at reversing. If there is sufficient work to keep a machine in use almost full-time then a self-powered unit is a better choice. Broadly, self-powered units fall into two categories, those consisting of two drums (or with a steerable front drum and two powered rear wheels that are also rollers) and those having one drum and two large pneumatic rear wheels. For the maintenance of established roads the all-drum type is very satisfactory, but where construction work is involved, the type having two rear tyres is the more suitable as it is better able to cope with steeper gradients and uncompacted soils. They are available with idle or powered drums and, where a lot of new formation compaction work is involved, the powered drum type has considerable merit, particularly in enabling the operator to counter side drift on cambers when the vibrator is in operation, as well as coping with slippery conditions.

Rollers weighing less than about 6 tonnes are not really practical for estate work and units weighing 10 to 12 tonnes are better. They will be more effective still if they are vibratory, i.e. the whole drum is made to vibrate up and down by means of an eccentric weight powered by the engine of the machine set inside the drum.

Besides smooth rollers which are best suited to compacting granular material, there are "sheep's foot" rollers, which have a knobbly surface designed to provide a much greater pressure over a much smaller area, and cage rollers which are formed of a very coarse square mesh grid of hardened steel and are used to crush stone in situ. Because of their limited application cage rollers are not likely to be of interest for estates. Sheep's foot rollers are very worthwhile having for road construction as they are more effective in compacting clay fills than a smooth roller and, used following a final grading on the earth surface before gravelling, provide a pock-marked surface that is ideal for keying the dressing into place. If the project is big enough to justify several rollers then one or two of them at least should be sheep's foot rollers. (See Pictures 2.22 to 2.24 on page 55.)

Rubber rollers which have two axles – each with several tyres on it – and are heavily weighted with water ballast, are not normally used for earth road work and are usually regarded as a machine for compacting asphalt layers. A rubber roller is not suitable for construction as a primary roller working alone although it can be used to follow a steel roller and improve upon its work. Indeed, when so used, rubber rollers are known in America as "proofing rollers" because they will promptly pick up and expose any inadequacies in work done by rigid steel rollers. As a maintenance tool working with a grader, a rubber roller has the following advantages:

- It can move very quickly and so keep up with a grader and compact every swathe as the grader works it.

- It has a degree of flexibility and so will compact the gravel that a grader spreads right down into a pothole. This is something a rigid roller cannot do: it is then necessary for a grader to rip and re-lay a potholed road surface to a depth greater than the depth of the potholes. The result may be the bringing to the surface of too much clay, followed by the need to redress the road with gravel to prevent it from becoming slippery during rain.
- It can operate very effectively on damp ground where a steel roller would "pick" and get itself hopelessly gummed up with a clayey medium.

The use of a rubber roller with a grader to maintain a busy main road where rain interference is a problem is a well proven practice in some countries. The practice deserves wider usage, especially where "little and often" is the maintenance method preferred.

Excavators

Most excavators nowadays are hydraulically operated. They may be mounted on tracks or on wheels. The smaller and cheaper units are often sold either as tools for fitting onto agricultural tractors or are agricultural tractors modified by companies specialising in this sort of product. The excavating mechanism is often complemented by a front end shovel which helps to give the unit some balance, enables it to prepare the ground it is to stand on whilst working, and greatly increases its ability to operate in and extricate itself from very muddy places. The better of these adaptions are very capable and versatile machines indeed. They will cope with the cutting of channels for laying culverts, roadside drainage (within limits) and, for a project which also has a building programme to do, will cope with the digging of foundations and the like. Being basically agricultural tractors they can be easily moved from place to place as may be required and a unit based on the estate's standard tractor can simplify spare parts problems.

Purpose built units – 37kW (50 hp) up to about 75kW (100 hp) – can be on either wheels or tracks but the biggest units – up to around 300kW (400 hp) – are all mounted on tracks. They are capable not only of large scale drainage but also of quarrying with both forward facing loading shovels and back hoes. All these machines are characterised by having great operating reach and a wide range of working heads: draining buckets of different shapes and sizes and rock buckets for tearing out quarry material; compressed air tools like drills, hydraulic rock breakers and piling hammers. Unless mobility between jobs is of great importance, the tracked units will be the most satisfactory for most uses. For estates with a substantial amount of drainage as well as a road programme to effect, one should be looking at machines in the 60kW (80 hp) to 100kW (135 hp) range (see Pictures 2.20/3.9 & 3.10 on pages 54 & 75).

Loaders

In the writer's experience, loaders made by adapting agricultural tractors are not satisfactory for road work. Purpose built machines, either articulated on wheels with permanent four-wheel-drive or on tracks, are necessary to do a proper job. The size range is vast, from about 15kW (20 hp) to 600kW (800 hp) on wheels and 30kW (40 hp) to 260kW (350 hp) on tracks.

Standard buckets only tip forward and for most needs this is adequate but side tipping buckets are available from some manufacturers. Also generally available are rock buckets which are much stronger than the standard item, skeleton, buckets which will pick up gravel and stone from a river bed whilst allowing the water to drain through, and multi-purpose buckets which have a movable cutting edge to allow the bucket to act as a blade or as a clamshell as well. The medium and large sized units can also be equipped with log forks for handling timber.

The choice of size for estate use will depend on a compromise between getting a machine manoeuvrable enough to be able to clear landslides and cope with other "small" jobs and yet big enough to load the project's tippers efficiently when earthfilling or gravel laying has to be done, and perhaps to tow a roller or trailer. (See Picture 2.4 on page 45.)

Picture Group 2:   Machinery

2.1   A lightweight bulldozer with A-blade and winch makes the best pioneer in steep and rocky country. In these situations agility is at a premium.

2.2   Never operate one machine in isolation. This machine broke through a layer of clay into an aquifer which filled the hole it was digging, to uproot the tree in the background, in minutes. The winch was under water and inaccessible. A second machine with a winch had to extract it.

2.3 A high-driving sprocket bulldozer equipped with ripper and S-blade. Compare with dozer in Picture 2.1 on page 44. Photograph by Señor Gustavo Sanic.

2.4 For projects in which money for capital equipment is limited to one machine, versatility is of paramount importance. The machine below will serve as a loader as well as a digger, has a good ripper and, where no timber haulage is involved, will do all the road-making work the project is likely to require.

2.5   Teamwork demands reliability. Here a light dozer cuts soil for filling into a dumper which is delivering to an embankment being built over a short swampy sector of road. The breakdown of any one machine renders the others idle.

2.6   The choice of optional extras is very important. This grader was able to dispose of a rock without straining the mouldboard or turntable and made it unnecessary to call in a loader or bulldozer for the job. The front mounted blade also allowed it to clean out runoff drains – of the type shown in Diagrams 28 and 29 on pages 117 & 118 – to the very end.

2.7   Choice of optional extras: the machine in the top picture is equipped with a rear mounted ripper, while the machine in Picture 2.8 has a scarifier mounted ahead of the mouldboard. The ripper is more robust and can rip right to the road edge where the scarifier cannot reach – an important point to consider.

2.8   Note the mouldboard blade in the lower picture has worn badly in the centre – see also Picture 2.9 on the next page.

2.9 The two halves of grader mouldboard blades should be switched from one side of the machine to the other every two or three weeks to even out the wear on them. They should not be allowed to become bowed like the one in this picture.

2.10 A grader windrowing freshly ripped hard clay prior to the addition of a dressing of crushed stone.

2.11 A grader with turntable-mounted blade positioned for grading the side of a road cutting.

2.12 Detail of the blade mounting of a grader without turntable. The yellow rams control height; the black ram, lower left and its counterpart (hidden) on the other side control angle; the small black ram (upper centre) controls the cutting angle of the blade – forward for windrowing and laid back for cutting and carrying.

2.13 The steering wheels must be able to turn sufficiently to more than compensate for the angle of articulation so that, when fully angled left as in this picture, the machine can still follow a right-hand bend. The ability to lift wheels off the ground with the hydraulics is useful for a quick wheel change (however, see text page 40).

2.14 Lightweight grader operating in straight position trimming road edge. Note the inadequate width of the ripper on this model. One more tine either side would enable the machine to rip right to the edge of the road.

2.15   Lightweight maintenance grader operating in the fully articulated position. It is much easier for the operator to control when angled like this.

2.16/2.17   Grader attachments to agricultural tractors are seldom effective but this unit was better than average and would be adequate for light construction and all the maintenance on a small project working ground that is not too rocky. It is shown here on maintenance work on heavy clay dressed with crushed stone and being prepared for a second dressing.

2.18   For the purposes of this book, this is termed a "dumper".

2.19   This is termed a "tipper".

2.20   Excavator fitted with a face shovel. See also Pictures 3.9/3.10 on page 75 for excavator fitted with back hoe.

2.21   Rubber rollers, usually used for laying asphalt, are also excellent for working with a grader doing maintenance. They are fast and can keep up with a grader, swathe by swathe, and will not crush the softer gravels to powder. They can also work damp soil when a steel roller would become clogged. Photograph courtesy of Baning.

2.22 Heavy vibrating padded roller for towing by a bulldozer. Not all padded rollers are this heavy. A simple padded roller can be made up from the front drum of a scrap road roller, upon which have been welded D-shaped pads made from $4 \times 1^{1}/_{2}$ inch steel bar and towable behind an ordinary agricultural two-wheel tractor. It can be a very effective tool.

2.23 Close up of the pads which can be of various shapes they do not have to be identical to these. Note the hooked tines spaced between the pads and which are placed both at the front and the back of the roller to prevent mud or roots getting packed between the pads.

2.24 The pock-marked surface left by a padded roller makes in very easy to key in a skinning of gravel afterwards.

2.25 Mobile sawmill. This unit has a horizontal bandsaw which travels through the log it is cutting. Light to tow and easy to operate, such units can produce lumber cut to very high standards. Photograph by James Andrews.

2.26 Four-wheel-drive tractor with logging plate. This sort of implement can be produced in the average estate workshop, is strong and simple to operate.

2.27/2.28 Four-wheel-drive tractor loading a log for sawing onto the dogging beams of a horizontal bandmill using a logging plate. Note the front mounted hydraulic winch. With this equipment the tractor is very useful for handling timbers for culverts and bridges. It is also a fine recovery vehicle for extracting bogged lorries or hauling in broken down equipment for repairs.

2.29 A horizontal bandmill cutting lumber for general construction. A second sawing station exists to the left, outside the picture, which is being loaded with another log as the log in the picture is being cut. For very long lengths, both stations can be used for the one cut.

2.30 Cutting lumber for bridge building. The timber is *Intsia bijuga*, hard, heavy and moderately durable. Picture taken from within the shed shown in Picture 2.27 on page 57 into which the mill is driven for shelter at nights. The simplicity of the foundation is self evident.

Dumpers and tippers

For the purposes of this book I will define tippers as being machines based on normal road lorry chassis and dumpers as being machines specially built for carrying rock, gravel or soil off sealed roads and therefore capable of working in much rougher conditions. There are lorries beefed up with four- or six-wheel-drive to cope with ill-made road surfaces and dumpers equipped with lorry-like steering, cab and controls, so there can be much confusion between the two types. However, let us take something like the ordinary two-wheel-drive, short wheel base lorry with a tipper body of from three to ten tonnes capacity as the archetypal tipper and the six-wheel-drive ten to twenty-five tonnes capacity, articulated steering vehicle as the archetypal dumper (see Pictures 2.18 & 2.19 on page 53).

For the delivery of gravel for road upkeep where the roads are of a reasonable standard and are not allowed to deteriorate to the stage at which four-wheel-drive is necessary, ordinary tippers are cheap to buy, fast running and, if the project has a tipper fleet for other purposes (oil palm fruit harvesting, for example), can bring with them the opportunity to standardise parts. They can be used for road construction too if the surfaces that they are to run over can be kept reasonably smooth and well compacted. Such simple tippers run into trouble quickly when ground is slippery, and even more so if it is soft. Their small front tyres will sink readily into uncompacted soil when they are loaded, and if the tipper happens to encounter such conditions unexpectedly at speed, the steering wheel can be wrenched from the driver's hand and the vehicle run out of control. The writer has seen several tippers overturned in this way. With careful drivers, and tippers equipped with four- or six-wheel-drive the problem can be ameliorated if not always cured, but these extras are costly and increase fuel consumption sharply.

Dumpers are much slower moving than tippers, are expensive to purchase, but are so strong and simple to operate that they are very cheap to run and, properly maintained, they last for many years. Their large off-road tyres and four- or six-wheel-drives enable them to traverse relatively soft and very muddy ground which a tipper would be quite unable to cope with. They can "snake" themselves out of difficult conditions using their articulated steering and are just about as unstoppable as any untracked machine can be.

The choice for a project must depend on the work to be done. If a lot of embankments have to be made that are too long to be built economically with bulldozers, then it is more efficient to cut, load and carry. If the ground to be traversed is clayey and rains can be a nuisance, then that may be a very good argument for obtaining a minimal fleet of dumpers, and the loader to load them, perhaps thereby reducing the number of bulldozers required. The choice may lie somewhere in between with either four- or six-wheel-drive lorries or the less specialised dumpers or, for a big project, in having some of each: dumpers in the early stages of development augmented by tippers later. If the delivery of crushed stone from a single quarry and crusher throughout the project is a major task then the use of fast running tippers is indicated. As with bulldozers, tippers and dumpers have a very hard life, so it is as well to choose and equip the units with care to ensure that they are working comfortably within their capacity. Most manufacturers offer rock bodies and cab shields for rock work; they are worth fitting. Dumpers are often available with tyre inflators as an optional extra. These are also worth having to save on loss of time when punctures occur. If several of the same model are to be operated, an emergency spare wheel is worth purchasing.

Four equal-wheel drive tractors (see Pictures 2.16/2.17, 2.26 & 2.27/2.28 on pages 52, 56 & 57)

On smaller projects where equipment is limited, four-equal-wheel-drive tractors – with the tractor's weight distribution two thirds on the front, one third on the rear wheels – can be extremely useful. Equipped with a powerful front-mounted hydraulic winch and rear-mounted logging plate, they can be used for marshalling logs and bridge materials, recovering other vehicles and hauling heavily loaded trailers, two or four wheeled. Not being tracked they can be dispatched speedily from job to job without damage to roads and have a tractive ability far beyond that of the equivalently powered two wheel drive tractor.

Crushers, drills and explosives (See Picture Group 9)

Unless there is no alternative source of road metal or there is a justifiable need on the project for a crushing plant for other purposes, the expense of crushing rock for road surfaces on a large scale is not something to be undertaken without a great deal of forethought. It is possible to get small mobile plants that can be towed by an agricultural tractor. These are very simple, consisting of an engine driving a jaw crusher on a frame mounted on wheels, the product dropping directly onto the road underneath. On small estates where there is a lot of coarse stone mixed in with the soil, the stones will tend to be thrown out to the road edges by graders during maintenance. To feed this material by hand through a mobile crusher, which is moved down the road by an agricultural tractor as the work proceeds, is a very good way of turning a nuisance into a blessing. It is unnecessary to screen grade the emerging stone provided that the jaws have been set to an acceptable gap allowing stone of say, $2^{1}/_{2}$ centimetres (1 inch) to fall through, the mix of coarse chippings and fines will combine well into the road surface when scarified in and spread by the grader.

The method noted above is fine for small scale work but for the serious production of large quantities of road metal a substantial deposit of raw material is required to justify the plant for a start. This can be river stone, either from the bed of a river which is bringing down large quantities of stone from its upper reaches or talus deposits in old, now dry lake beds, as examples of two possible sources. Otherwise it is likely that massive rock will have to be tackled, usually with the use of explosives. If there is a choice of rocks in the project concession it is better to use medium to coarse grained volcanics or hard limestones and sandstones that break down into blocky pieces, than to select fine grained hard or glassy volcanics, flints or cherts, all of which will break into very sharp-edged shards. The sharp edges can do a lot of damage to vehicle tyres until they have rounded off with usage. The very hard fine grained or amorphous rocks also tend to to be much harder on the drills, loaders, dumper bodies and crusher jaws than the more coarsely structured stones.

Although other forms of crusher exist, it is most likely that a jaw crusher would be chosen for estate work. The smallest units produce about 4 tonnes per hour and the range extends upwards well beyond anything likely to be of use to an estate. The crushing unit and engine form the nucleus of the plant which may be augmented by an automatic feed into the jaws, and conveyors and screens on the output side to separate the various different sizes of stone produced. For earth roads there is no need to screen the product provided the jaws are set to allow no stones of more than 5 centimetres (2 inches) thick to fall through. This will in fact allow stones that may be wider or longer than this to pass through but most stones will not, in length or width, exceed their thickness by more than about 50%. Stones larger than this size will be difficult for a grader to spread smoothly during the work of dressing a road surface so this should be regarded as the upper limit. Use closer settings, down to $2^{1}/_{2}$ centimetres (1 inch) if possible. Screens will then be unnecessary and this can save a lot of money both in purchase and running costs because screens wear quite rapidly. If the crusher cannot be tuned to produce stone that small, then a screen does become necessary to separate out all the stone bigger than the largest acceptable size. This can be recycled through the crusher, preferably mixed with fresh material. Alternatively, the coarse material can be reserved for special uses where the bigger stone is required.

It is suggested that a unit consisting of an automatic feed, jaw crusher and engine, screen if necessary, all mounted on a unitary base should be aimed for. The base can be shaped like a sled or form a chassis with rubber tyred wheels (equipped with jacks to take the load off the wheels when the crusher is operational). Such a base will remove the risk of the main components becoming misaligned with settlement and allow the project the flexibility of moving from one quarry to another without too much time wasted in dismantling or reassembly if it is required to change the source of raw material. It is further suggested that the unit be sited on a stepped site, the top level being used to stockpile uncrushed stone, the lower level being the place for the crushing plant (the alternative is to use a heavy duty conveyor belt to feed the crusher – additional expense, and one more thing that can break down). The difference in height between the two levels should equal the height of the crushing unit from the bottom of the sled runners to the level of the entrance of the automatic feed plus about $^{1}/_{2}$ metre (20 inches). An inclined chute can then be placed at the edge of the higher ground level to receive and feed uncrushed stone into the automatic feed to the crusher. The crushed stone can be dropped onto a portable conveyor belt for distribution to either a hopper for loading the tippers, or a simple stockpiling yard from which the tippers can be loaded by a front end loader. Unless the scale of operation is very large the latter method is preferred. One front end loader attached to the plant can load the uncrushed stone directly into the shute for crushing and load tippers as they come to collect the crushed stone. The face of the site where the crusher nestles under the upper level will need to be supported and protected by revetting, gabions or some other means to prevent it eroding or collapsing and act as the support for the upper end of the chute.

In selecting a crusher, make sure that the rated output is more than adequate and that the jaws have the capacity to receive the size of stone that is available. Having to break down the raw material by hand or with jack hammers or rockbreakers to an acceptable size introduces extra costs which can be quite heavy. Where the use of explosives is involved the skill of the blaster can be crucial. By utilising the right techniques to get a good "shatter" when the rock is blown up, the blaster can reduce this problem to a minimum – little more than the cracking of the odd rock in half using a sledge hammer with a rotan handle. Reliability and resistance to wear are points of prime importance when choosing between makes and in this the metallurgical skills of the manufacturer are paramount. Poor quality jaw liners can wear out in a matter of hours and the strains put on the eccentric shaft and bearings and on the body of the crusher itself are enormous. It will pay the intending purchaser to enquire from other quarry operators as to what machinery has proven to be good in the region before making a decision. Be very sceptical of the claims of the salesman! There are a lot of very second rate pieces of equipment offered for sale in this field.

There are two or three premier manufacturers of compressed air rock drilling equipment in the world and their equipment is reliable and effective and can be bought with confidence. The most popular compressors are units that are mobile to the extent that they are mounted on a chassis with towbar and pneumatic tyred wheels, suitable for towing behind a four-wheel-drive pick-up or a mobile drilling rig. This is useful not only in the quarry but enables the unit to be used for the drilling and blasting of rock outcrops that may be encountered during the process of earthmoving for road making. However, do make sure that the compressor chosen is more than just theoretically of adequate capacity – disappointment is nearly always the result of inadequate compressor capacity for the type or number of drills that the machine is eventually expected to run simultaneously.

Most operations will start with hand held jackhammer drills. These can be powered by compressed air or by an integral internal combustion engine. The latter type is only suitable for shallow holes as compressed air is needed to clear the deeper holes of dust. Manually operated drills will only drill small holes, usually 32 millimetres (1¼ inches) in diameter. This limits one to the use of the more dangerous high explosives rather than the safer emulsions and slurries, and implies the use of more holes than would be required if larger hole diameters could be used.

Once started, a quarry will be better off using mobile drilling rigs sometimes called wagon drills (see Picture 9.5 on page 232). They will drill 102 millimetre (4 inches) diameter holes which are beyond the capability of manually operated machines and which are far more suitable for serious quarrying work.

The choice of explosives, fuses and detonators will depend on the type of rocks to be blown up, the size of the operation and local availabilities and regulations. Explosive types, hole sizes and spacings are interrelated and must be considered as a package, each item complementing the others and suited to the rock type to be tackled. The objective should be to shatter the rock in as inexpensive a way as possible without throwing it around any more than can be helped and creating as little noise as possible. Modern explosives, properly used, are very effective at this (see the section on "Quarrying").

Sawmilling equipment (see pictures on pages 56 to 58)

Where durable timber is plentiful it is likely that the Project Manager will consider the acquisition of sawmilling equipment for the provision of timber for buildings, bridge timbers and decking, fencing etc. Mobile sawmills with circular and band saws, both horizontal and vertical, are available nowadays (see Picture 2.25 on page 56). They are practical, easy to move from place to place and to use. Simple equipment for automatic sharpening of the bands is also obtainable for some makes and models. Their one drawback is that the length of log that can be sawn is limited. Where long beams are not required this is no problem. However, some projects may need to cut long beams and capsils for major bridges and have suitable, durable logs available for the purpose. In such a situation some thought might well be given to the use of a horizontal bandsaw as the primary breakdown unit for the mill. Horizontal bandsaws travel on rails as they cut through the log, rather than the log being sent past the bandsaw by a cradle. They have three clear advantages over conventional mills:

- The saw can cut any length, the only limits being the length of rail line for it to travel on and the length of log available for cutting: this enables bridge beams and capsils to be cut level, top and bottom, as long as may be required and in one piece.
- The cost of such a mill is lower and it is easier to install than a normal vertical bandmill unit of the same capacity because there is no log carriage required that must be designed to take the very largest logs to be handled.
- The mill can be moved with the loss of no more than the foundations holding the rail lines – these can be of wood or of concrete but either way are simple and inexpensive.

Bandmills that are properly looked after cut their timber far more accurately and with far less sawdust waste than circular saws because their kerfs are much finer. With modern bands and sharpening equipment they are no more difficult to maintain than circular saws.

A horizontal bandmill can be a complete mill on its own or, to increase production rates, it can be backed by an edger and band resaws. There is therefore a great degree of basic flexibility in the concept which should appeal to the practical estate manager.

Beside sawmills, when working in forest it is necessary to equip the road teams with large chainsaws capable of felling and cutting to length any trees that may be encountered – this both to ease the work of the bulldozers and to make possible the reclamation of timber that would otherwise be destroyed.

In areas that have been fought over, bullets and shrapnel embedded in trees can be a hazard for all forms of sawmilling equipment. It is wise in such circumstances to obtain an effective metal detector and check all logs prior to milling. The cost of the metal detector will be recovered if it saves just a couple of saw bands.

GETTING THE RIGHT BALANCE

In some senses this is an extension of the thinking behind ensuring that the individual machines are equipped with the right optional extras. The nature and quantity of work to be done and the time in which it must be completed (with due allowance made for weather interference) must be considered and an assessment made of when and what types and quantities of machines will be required in order to achieve the desired results.

Taking as an example an area consisting of several low hills with a number of alluvial valleys between them, it could well be that no continuous road formation work can really begin until extensive drainage has been effected, both to drain the broad low lying valleys for the crops to be planted, and to enable the ground to firm up enough for the road work to begin. In such a situation the initial decision could be to bring in a couple of tracked backhoes and a couple of bulldozers of about 60kW (80 hp) equipped with winches and mounted on broad tracks. The bulldozers would help the backhoes to cross hills between valleys and make temporary roads for the supplying of diesel and provision of service to them. The backhoes should firstly create or clean out all the main drainage lines, secondly form the roadside drains in the alluvial areas and finally tackle the field drainage prior to planting. The small bulldozers could start some of the lighter permanent road work on the hills. Later on, when sufficient land has been drained, larger bulldozers equipped with rippers and complemented by a loader and two or three dumpers can be brought in to start the major cuttings through the hills, whilst simultaneously providing fill for the embankments across the valleys. The small bulldozers would then be used to spread and form the fill.

A bridge building and culvert laying capability will have to be developed. By this time compactors and a grader will be required and finished roads will begin to be produced. If there is no gravel available but stone has been found there will be the need to set up a quarry and crushing plant with a loader and, with either the existing dumpers or with additional tippers, to deliver the crushed stone to the roads for surfacing. As time goes on and the drainage work is completed one or both of the backhoes may become surplus or can be diverted to culvert laying and quarrying. As the system nears completion some bulldozers will become surplus and the accent will move to maintenance with graders, rollers, tippers and loaders becoming the dominant requirement. During the whole of this process it is essential to ensure that the machines complement one another. Bulldozers should not be expected to work in inadequately drained land. The roles of the pioneer bulldozers that open a trace and the heavy bulldozers that cut and form it should be appreciated. The compactors that consolidate, the graders that finish off the formation and make the upper parts of the drainage system, the loaders and tippers that provide the gravel, the grader that spreads it, the roller that firms up the running surface, should all work together and one should not be kept waiting for the other the whole time. It is a question of attempting to achieve and maintain the right balance overall to keep the machines as consistently productive as possible.

Temporary imbalances will occur. For example, the grader may be underworked in the early stages – a good machine with a first class operator is capable of an immense amount of work and can be switched quickly from place to place so that it may easily cope with all the work that a couple of bulldozer teams can throw at it initially. Later, when a considerable length of road has been completed and is in use, maintenance will begin to occupy more and more of the grader's time until it is always busy. This change in the grader's work load may take a year or more to come about so, if the grader is idle part of the time in the beginning, this is no reason to believe that too good and powerful a machine has been bought. To ameliorate this effect, bulldozer operators can initially be encouraged to leave as much of the shaping up of the formation to the grader as is possible. Later, when the grader is rather busy, skilled dozer operators can do much of the work of cambering, super-elevation and cutting of side drains and their outlets, leaving the grader just the finishing touches to do.

Some of the work can be quantified. For example the thickness of gravel to be spread over the roads multiplied by the width of the running surface multiplied again by the total length of the road system can give a gross figure of the amount to be quarried and moved. This can be broken down by years relating to the development plan and supplemented by the estimated annual replacement of gravel required for maintenance (say ten per cent) to give a picture of annual gravel requirements. Depending on the siting of gravel quarries, average delivery distances can be worked out to give total tipper miles that have to be run to effect the delivery, and from that the size of tipper fleet required. From this also can be estimated the number of loaders required. If crushed stone is to be used this information will help in the selection of crusher capacity, number of drills, size of compressor and consumption of explosives etc. If the project has been well mapped, calculations for the equipment required for making road embankments can be made in the same way. However, it is as well to remember that these calculations will give "ball-park" figures that will be useful as a guide, they should not be taken too literally because there will always be the unexpected to reckon with, sometimes favourable, sometimes not. But at least this sort of exercise will help to ensure an effective and commercially acceptable balance of one's equipment.

FORMING TEAMS

Large estate projects where several hundred kilometres of road have to be built over a period of a few years will probably require more than one road team to do the job. The heart of the road team is usually a group of bulldozers. These machines should be selected to cope with the topography, vegetation and soil that they will encounter. Where steep primary or secondary forest land constitutes a significant proportion of the ground to be tackled, a team might consist of:

- Two machines of around 120kW (160 hp) with winches, one with an A-blade and the other with an S-blade and hydraulic tilt.
- Two machines of around 170kW (230 hp), one with a winch, the other with a ripper and both equipped with S-blades and hydraulic tilts.
- One machine of around 240kW (320 hp) with a ripper and an S-blade with hydraulic tilt.

This mix will allow one of the light machines (with the A-blade) to be the pioneer on steep slopes, cutting the initial ledge upon which the bigger machines can subsequently work safely, and scaling foothills where high cuts have to be made. The other machine would do a lot of the formation finishing work including the cutting of side drains that a grader can follow and, if needs be, cambering and super-elevation. These light machines will do all the cutting of culvert foundations too big to be effected by backhoe and will, with the middleweights, "close-off" work at night. With their winches they will be able to get each other out of trouble and cope with the movement or recovery of logs for milling, culverts or bridges. The middleweights will do much of the earthmoving following the initial work of the pioneers. The one big winch will tackle trees too big for the lightweights and will be available for the more difficult recovery problems. The ripper on the other middleweight will greatly speed up work on indurated tropical clays and soft rock. The one heavyweight, if such a machine is required, will be the bulk earth mover to cope with all the deep cuttings, fell the biggest trees, dig out the biggest stumps and rocks and tackle the really hard ripping work.

It will be obvious from this that each machine has its place and that on hillsides the small machines will lead and finish off work whilst the larger units concentrate on the heavy slog of shifting soil. On flatter forested ground a different team might be more effective consisting of:

- One machine of around 120kW (160 hp) equipped with a winch and an S-blade with hydraulic tilt.
- One machine of around 170kW (230 hp) equipped with a ripper and an S-blade with hydraulic tilt.
- Two machines of around 240kW (320 hp) equipped with S-blades and hydraulic tilts, one with a winch, the other with a ripper.

In this situation the winch equipped heavyweight would be likely to lead by clearing the forest. The other, with the ripper equipped middleweight, would perform the majority of the earth-work whilst the lightweight would cut culvert sites, place culvert logs, move bridge timbers, cut side drains and prepare the formation for the grader.

These two examples are given to indicate that teams of bulldozers should be formed of machines carefully selected and equipped to complement one another. There are, of course, other permutations that can be reasoned through to fit other circumstances. Bulldozers are very expensive and it is essential to choose and match them well to get the very best out of them.

Teams do not have to be as big as five or six machines. Two units, well matched for a purpose, can constitute a team. Neither need a team consist of bulldozers alone – other plant (compactors for example) may well be involved in the work at this stage but graders are unlikely to be fully occupied if attached to a construction team. It is better that the grader is sent as required, occupying itself with maintenance between times. Another sort of team that might be formed for estate road building is a cut-and-fill team consisting perhaps of a face shovel or backhoe, or front end loader and a bulldozer, with a group of dumpers or tippers delivering fill to an embankment where a light bulldozer and compactor are forming the road. A road maintenance team can be formed around a grader with appropriate equipment in the same way.

In the writer's experience, one of the benefits of team forming is that the team members soon develop a degree of camaraderie and with it a great pride in their work – woe betide the team member who is below standard or the new boy who does not make a genuine effort to achieve. Such a group is easily motivated by perceptive management and be a real pleasure to work with. A team leader is usually necessary though he may be no more than a senior operator working one of the machines, respected by the others for his skill with the machines and by the management for his ability to teach and inspire. Supervision should be given by the road foreman or management, and support is as important as supervision. By that is meant transport out in the morning and back at night arriving on time, without fail. Prompt transport back if rain makes work impossible; seeing each machine started up first thing in the morning before leaving the team to work; quick reaction to breakdowns and accidents and convincing the team that management is very interested in what the team is doing, recognises its achievements, helps with its problems and reliably turns up on site several times a day. With such a system of co-operation it is then possible to ensure that machines are refuelled, engines are properly run down before they are shut off, that tracks are cleaned out and that the work is correctly closed off against the possibility of rain each time work stops. Such a team will usually work hard all day long and the output can be immense. Bearing in mind that the cost of the operator is very little compared with the depreciation and running costs of the machine, it has been the writer's practice to employ two operators per unit working a twelve hour day during which the machine never stops, the operators taking alternate turns at about two hour intervals. Not only does this mean that the operators are always fresh and alert, and therefore output stays up, but it ensures that there is always another operator on hand to help immediately in the event of an accident. With team operations, operators usually cover one another when someone is ill or absent for some other officially acceptable reason. The net result is to maintain the pressure of work at its maximum all the time.

MACHINE MAINTENANCE

Most earthmoving machines are fitted with meters on the engine which record "hours" worked. In fact they are a record of engine revolutions in some and of fuel expended in others. They will not be true measurers of time (and should NOT be used for pay calculations) but are effective as a means of assessing when the machine is due for maintenance. The schedules will be laid down in the machine manuals and should be adhered to, otherwise not only may damage result but manufacturers' guarantees may be rendered null and void. Some manufacturers operate a back-up service that includes being prepared to service their machines for a charge. In many cases this does not apply and the service will have to be done by the project workshop. The majority of earth moving machines nowadays are easily serviced in the field and, for this purpose, the estate will need to have a service vehicle which can carry in it all the tools, oils and spares required for the job. Even bulldozer track changes are so straightforward that they can be done by the roadside. The vehicle used should therefore be sturdy and able to deliver quite heavy spare parts. Such work can be done at night if the workshop staff are geared up to that sort of operation and this will improve the machines' availability for work during daylight.

Operators should check around their machines every morning before starting work and again before shutting off and cleaning down for the night. The road foreman and manager should also do a quick check over at least once a day to look for things like A-frame fractures, seized rollers, worn cutting edges, loose bolts and other defects that the operators may not notice while they are working. Listen also for slight changes in the sound of a machine at work – a change in note may be the first sign of trouble with the turbocharger, for example. Often it will be the visitor rather than the operator who will note that sort of indicator – being on the machine the whole day the operator will be too used to the sound of it to detect a gradual change as easily as someone fresh to the machine. Drive the machine once in a while and see that all the controls feel right – again, a gradual deterioration of adjustment will go unnoticed by the operator until it is quite bad, whereas a stranger will notice it immediately. Catching these sort of defects before they have become serious can prevent accidents and reduce repair costs and unwelcome down time.

Each machine's servicing and repair history with costs should be entered up regularly at the workshop, both to ensure that the work is done properly and as a management tool to enable machine efficiency to be assessed.

Chapter 4

Design and construction

OBJECTIVES

The main objectives of road design are:

- To provide a safe running surface for the vehicles that will use it.
- To make the road resistant to wear caused by traffic and to erosion caused by the weather.
- To achieve these aims satisfactorily at minimum cost.

THE FORMATION

Basic principles

In some senses a road is a bit like a house – it needs firm foundations, a roof and the necessary guttering and piping to shed the rain and keep the core of the structure dry and unspoiled. The roof of a road is provided by the camber, or by super-elevation on corners, and the drains and culverts take the water away from the structure. The camber and super-elevation for a sealed road need only be very slight in order to shed the rain but on an earth road they have to be exaggerated in order to be effective. This is partly because an unsealed surface tends to be rougher than a sealed surface – this in itself impedes runoff – but more importantly, because the unsealed road can soon absorb enough of the rain water to suffer significant damage and therefore, the more rapidly the water can be got rid of the better. It is always wise to make the camber rather high to start with, firstly because it will settle in time anyway, and secondly because it is easy to cut down a high camber that is firm but difficult and expensive to build up a camber that has proven inadequate and degenerated into a muddy morass.

A camber that is too high will also bring problems in use. Vehicles will tend to "track" in the centre of the road in order not to be tipped to an uncomfortable angle at which the driver will feel insecure. This will localise wear to two narrow strips either side of the crown of the camber which will deteriorate prematurely and require that maintenance rounds are effected more frequently.

To some extent resolution of the question of how high a camber is right will depend firstly upon the natural qualities of the soils of which the road is built and the surface dressings used, and secondly upon the steepness of the gradient: the steeper the gradient the higher the camber must be to ensure that the water is shed to the side and does not flow down the running surface of the road (see Picture 1.2 on page 22). There is therefore no point in specifying a set camber for a road. When new, as noted above, the road will need a higher camber than when it has been in use for several years. By then the long term effect of induration and compaction by traffic, and the accumulation of several years' worth of supplementary dressings of gravel during maintenance, will have made it far more resistant to the damaging effects of rain. Observation of the road, and constructive reaction to what is observed by the person responsible for the road, will enable cambers and super-elevations to be achieved that are suitable for the conditions and circumstances of each individual sector of the road. To get the best results at the minimal cost, regard the making and maintaining of such earth roads as a craft not a precise science. The essentials of being a good craftsman are constant personal attention to the job in hand, skill in the use of the tools and the ability to learn rapidly, by experience, how best to get good results with the materials to hand.

In the wet tropics, whenever there is a risk of rain falling, one can take precautions to mitigate the effects of a downpour by:

● Not stripping the ground of vegetation beyond what can be formed before the rain falls, certainly not beyond what can be shaped in a day.
● Ensuring that work is always well compacted as it proceeds.
● "Closing off" the work properly each evening. By that is meant roughly cambering the work with the bulldozers; rolling it smooth, if possible, to ensure that if rain comes it cannot lay overnight and soak into uncompacted soil; making sure that all the roughly formed drains are open and that the water can get right off the formation unimpeded.

By so doing time lost with the earthmovers following a spell of wet weather will be minimised.

Because the work in hand is on a project where the plantable soil is the project's main asset, the earthmovers should not be allowed to wander where they please around the work site. Some damage to the plantable soil is inevitable but it can be kept to insignificant levels by good management and good working practices. If felling, burning and clearing are to follow on the land alongside the road the vegetation can be pushed into the field for destruction with the main burn. If not, or if one is putting the road through existing crops, the trees felled, if any, should be cut into manageable lengths as quickly as possible to enable the machines to stack the brash neatly where it will be least nuisance. This work is dangerous, trees felled and pushed by bulldozers will probably experience bending stresses and can snap and fly when being cut: very skilled chainsawyers or axemen are necessary for this work.

Whilst all this is in progress the survey markers of the original trace will be knocked out and lost. A small gang of two or three skilled labourers with survey gang experience will be needed to continually replace the sighting rods on straight sections of road when the vegetation has been cleared and before the shaping of the formation commences. They should at this stage

Picture Group 3: Construction

3.1 A light bulldozer equipped with an S-blade and a hydraulic tilt quickly closes off work for the night with a coarse camber to shed any overnight rain. The road is a harvesting road on a cocoa estate. Road construction preceded logging, land preparation and planting.

3.2 A harvesting road ready for grading. Note the sighting rods to keep the roadwork straight and neat. The left hand rods are on the original survey trace.

3.3 Main road construction: forest clearance was started early in the morning, this picture was taken at about ten o'clock.

3.4 The same sector of road pictured at about two o'clock. By nightfall it had been graded and rolled to leave a free draining surface and the bulldozers were ready to tackle the next stretch the following morning. Gravelling with crushed stone was effected two or three days later after overnight rain. The local soil consisted of a well structured red-brown clay with excellent bearing qualities.

3.5 Ripping the freshly dozed trace to pull out roots and rock before grading is always a wise precaution and can save damaging the mouldboard and turntable. The road is a main harvesting road for cocoa.

3.6 The use of the front mounted blade to knock the tops off piles of rough dumped gravel, whilst simultaneously rough spreading with the mouldboard, is demonstrated by this grader.

3.7 Laying corduroy over a section of coastal swamp: note the piles of mud either side of the road which will help seal the logs from the atmosphere after covering. Between half and one metre of stoney red clay was laid over the logs, compacted and dressed with crushed stone to take forty tonne lorry loads.

3.8 Filling over a soft valley with dumpers, grader and roller. Continuous delivery, spreading and compaction ensure a satisfactory embankment and minimise rain interference.

3.9/3.10 The use of a back hoe for drainage: broad round-bottomed drains like this are resistant to slumping and last a long time between cleanings. The long reach of the machine is what enables it to cut broad drains so well. In this case it was able to place the spoil right across on the other side of the road where it was left to harden and later used to raise the road level. This put the road surface above flood level.

3.11   Laying raised coral on a cocoa estate main road under construction. The coral does not mix well with the underlying clay and has to be laid as a thick blanket cover.

3.12   Coral will provide an almost cement-like surface which is very durable if well rolled down with water. Sufficient water should be applied to make it damp but not slushy.

3.13 Spreading "rotten rock": the material – weathered rock not hard enough to crush – was taken directly from the quarry and dumped on the road by a fleet of tippers. The bulldozer both rough spread the material and crushed the large lumps under its tracks. A few large and very hard pieces were separated out and thrown to one side but little was lost in this way.

3.14 Subsequently the road was rolled and used in that condition for a few weeks before it was ripped, graded and rolled to a good smooth finish.

3.15   Upgrading a main road on a cocoa estate: taking good soil from the inside of a corner to improve visibility and spreading it over an embankment to provide a good seedbed for cover crops to stabilise it.

3.16   This was part of an operation involving widening, radiusing of corners and preparing the road to become the main access route for the project's factory.

3.17   Hairpin bends: the road drops down at 1 in 15 to the left and rises at 1 in 15 behind the vehicle. On the bend itself an incline of approximately 1 in 30 ensures trouble free drainage. The post on the extreme right of the picture is the focus of the radius and the other markers are a guide for the grader both at the time of construction and for later maintenance. This is part of a main road connecting a cocoa estate to its wharf over a 350 metre high hill.

3.18   The point where the road is almost undercutting itself, to the left of the vehicle in the upper picture, sometimes needs revetting if there is risk of the upper section falling away. Such a revetment is shown in the lower picture. It consists of hardwood posts dug one metre into the ground backed by hardwood baulks just laid in place and backed by fill as the height of the revetment increases. When finished, the revetment was filled to the level of the tops of the posts. No bolts or nails were used in the process.

measure in the inner radii of corners and it is suggested that these be marked with short thick little posts hammered well into the ground so that they will not easily be knocked over by the machines as work proceeds (see Picture 3.17 on page 79). To maintain really neat, straight roads, short posts can be used on the straights when the bulldozers have finished and the grader follows up (see Pictures 3.1 & 3.2 on page 71).

A few tips and ideas:

- It will pay to rip after clearing forest or tree crops to get all the roots out before the shaping of the formation begins; not only will this ease the work of the dozers and of the grader subsequently, it will also prevent the markers being pushed over when roots are torn out as the earthmoving proceeds.
- When cutting a series of parallel harvesting roads, use the trace lines at one or other side of the actual road rather than the centre. This will save a lot of remeasuring and hassle trying to re-establish the trace markers. However, make a clear decision that within any one field all the roads will be on one side or the other, or differences in block widths between roads will arise.
- Always put the markers a metre outside the road edge otherwise the bulldozers will be constantly knocking them out as they make their passes to cut the roadside drains. Thus if the harvesting road width is to be 5 metres, then put the markers in 7 metres apart so that the machines have a metre clearance either side to work to within the trace markers.
- Regard the trace markers on a hillside road following a set gradient as indicating to the dozer operators the level of the finished road surface and not the edge of the road. This is something they may find hard to understand until they are used to it. These markers too will be knocked out and it will be necessary for the management or the surveyor to check the road frequently to ensure that the gradient is being followed reasonably accurately.

Profiles for straight roads

Flat ground

The shape of the formation will be affected by the proposed road's designation and usage. Harvesting roads should be superficial and occupy as little land as possible to be consistent with the needs to allow vehicles to pick up loads of crop, and in emergency to pass one another, whilst still conserving plantable land. A main road will not necessarily be superficial because the alignment must allow for higher travelling speeds and the need for vehicles to pass in opposite directions or overtake one another without having to slow down unnecessarily. Substantial cuttings and embankments may therefore be necessary along the length of the road to gain an acceptably even profile. There is also the need to provide vision between the edge of the planting and the edge of the road so that anyone emerging from the plantings has time and space to see and be seen by vehicles before stepping into the road. Being wider than the harvesting road, more water will be shed from the main road during heavy rain so the side drains for a main road should be more substantial. A straight harvesting road profile is shown in Diagram 5. In Diagram 6 a straight main road profile is shown. Both assume flat ground on either side.

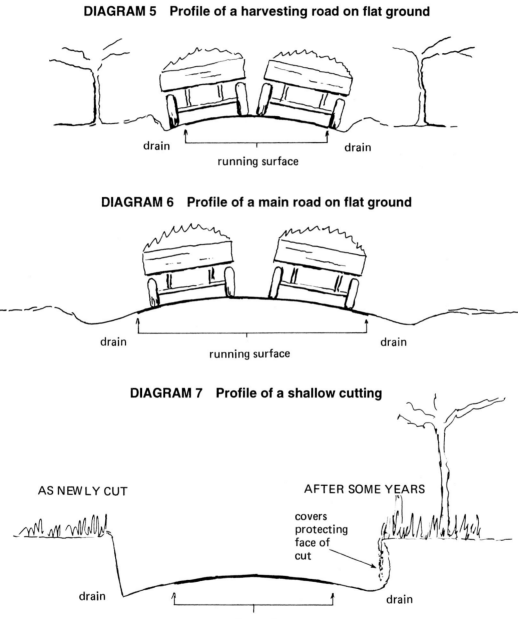

DIAGRAM 5   Profile of a harvesting road on flat ground

drain

drain

running surface

DIAGRAM 6   Profile of a main road on flat ground

drain

drain

running surface

DIAGRAM 7   Profile of a shallow cutting

AS NEWLY CUT

AFTER SOME YEARS

covers
protecting
face of
cut

drain

drain

running surface

DIAGRAM 8a   Profile of a deep cutting

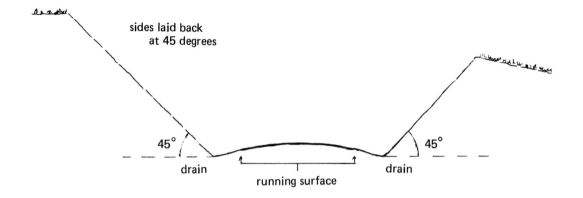

sides laid back
at 45 degrees

45°   45°

drain   drain

running surface

DIAGRAM 8b   Profile of a deep stepped cutting

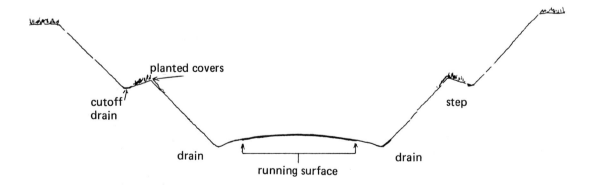

planted covers

cutoff
drain   step

drain   drain

running surface

DIAGRAM 8c   Perspective sketch of a deep stepped cutting

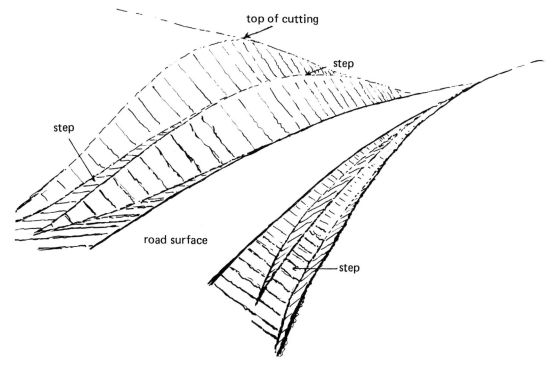

DIAGRAM 8d   Profile of a deep cutting in inclined sedimentary rock

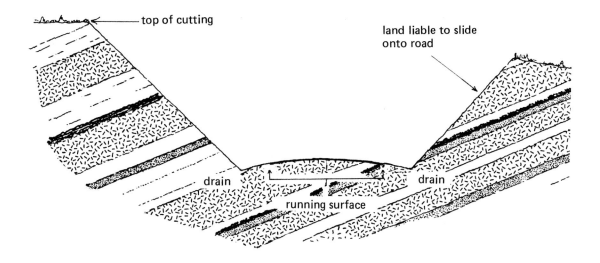

Note that on a harvesting road there is only need to provide for just enough room for two of the widest of the vehicles likely to use it to pass one another; normally the vehicles will drive in the centre of the road. The planted crop can be as near as one metre from the road edge. For main roads there is need to provide for a metre clearance between the widest vehicles and another metre of clean running surface either side of them. The nearest planted crop should be three or more metres from the edge of the running surface, substantially more on the insides of corners and on the lead in to corners and junctions. This extra width is all for safety purposes.

Shallow cuttings

A profile for a shallow cutting applicable to main or harvesting roads (up to two metres deep at the most) is shown in Diagram 7 on page 81. It is not usually necessary to slope the banks of shallow cuttings in good soils. Near vertical banks will stand up quite well. However, provide extra width at either side, beyond the normal position of the drains, to allow for some falling of the soil from the sides to occur over time without it being able either to hamper drainage or pinch the road width. Do not destroy the grass or other herbaceous growth on top of the cutting – indeed, if necessary, sow additional seed or plant more seedlings or cuttings of suitable cover crops there as an anti-erosion measure. Tree crops can be planted quite near the edge and will help secure the ground. On the left hand side of Diagram 7 the profile is shown as it will be when fresh cut and on the right side as it will be after a few years. Note that the vegetation will by then be sheltering the sides of the cutting from the direct effects of heavy rain drops falling vertically onto it and this umbrella effect should be encouraged. If gradients are slight, take great care during the earth moving work to ensure that a continuous fall is maintained – a hollow in a cutting will fill with water when it rains and become a morass of mud that is difficult to cure.

Deep cuttings

A profile for deep cuttings applicable to main roads is shown in Diagram 8a on page 82. The sides should be sloped at an angle at which the soil is not likely to break off and fall into the road. Before earthmoving starts the top edge of the cutting should be marked out to allow for the slope (if a 45 degree slope is used, then the cutting width will be the total width of the road formation plus twice the depth of the cut). In practice it will be found that skilled bulldozer drivers will instinctively allow for this adequately. If the cutting is very deep the sides may have to be stepped as in Diagrams 8b and 8c on pages 82 & 83. The stepping is intended to carry water during heavy rain from the level above the step off the formation before it can reach the road. To do this, the step must be wide enough and the tread sufficiently super-elevated away from the road to ensure that, despite the silting effect produced by soil washed off from the face above that will occur during heavy rainfall, subsequent drying out and crumbling, the water does not overtop the outer edge and flow down the face below onto the road. The step is in effect a perched cutoff drain. Where the soil or rotten rock being cut through is not homogeneous but consists of alternating sedimentary layers of softer and harder material, it may be very difficult to maintain the step in an effective condition. Despite this, the back of the step at the foot of the upper face must be kept clear like a drain until it has been possible to establish adequate vegetation on its outer edge to reduce erosion risks to acceptable levels. If this cannot be done then it is better not to waste time stepping the slope but to lay back the slope of the sides as far as is practical. Leave sufficient room to cope with any normal off-wash either side of the road formation and accept that one will have to attend to any slides that occur, as they occur. Concentrate on planting covercrops to protect the bare soil from the destructive effects of heavy falls of rain as soon as possible.

When working in sedimentaries keep an eye on the strike and dip of the rock. A strike parallel to the road and a dip towards the road may result in massive landslips along a line of cleavage exposed as a result of the cutting (see Diagram 8d on page 83). If the bedding is more nearly horizontal than shown in Diagram 8d, there may still be spring lines where water flows between the strata to emerge in the cutting, these too can eventually trigger off landslides.

The method of making the cutting will depend upon the material being cut and the distance it has to be carried. If the distance is not great – no more than 100 metres say – then bulldozers with S-blades and rippers will be the first choice. If the material has to be carried for two or three kilometres then scrapers may be used (but see Chapter 3, *Tools of the Trade* page 38 for the reservations on the use of these machines). If scrapers are unsuitable or not available and the material has to be carried for anything beyond the economical limits of dozing, then use a cut-and-fill team (see Picture 2.5 on page 46).

Side cuttings

Two profiles for side-cutting are shown in Diagrams 9a and 9b on page 86. In Diagram 9a a profile is shown which has the crown of the camber two thirds of the width of the running surface away from the face of the hill. This reduces the amount of water which is shed to the edge of the road under the face of the hill, compared with the profile shown in Diagram 9b, in which the whole road width is super-elevated to slope away from the outer edge of the road. A normally balanced camber on a side-cutting is not likely to be popular with drivers because those driving on the outer side of the road will always be leaning towards the drop below. In the profile of Diagram 9a, vehicles travelling on the outer half of the camber will be more or less upright and additional comfort is provided by a kerb on the outer edge consisting of some of the soil thrown off the cut by the bulldozers that made it. Whilst such a kerb would not prevent a speeding or very big vehicle from going over the edge if it got out of control, its very softness will enable it to hold a smaller vehicle travelling at a more sensible speed with a good chance of preventing it from going right over the edge. The drawback is the need to drain off water over the edge in such a way that there is never enough discharged at any one place to cause gulleying of the fill. This is best effected by hand cutting small (say twenty centimetres wide) outlet drains through the kerb every two or three metres, at right angles to the road alignment, as indicated by the dotted line in Diagram 9a (see also foreground Diagram 21 on page 100). To use this profile without a kerb looks pretty frightening to a driver on the outer edge and, indeed, can be dangerous. If hand maintenance is not acceptable, then to facilitate maintenance

DIAGRAM 9a   Profile of a side cutting, cambered

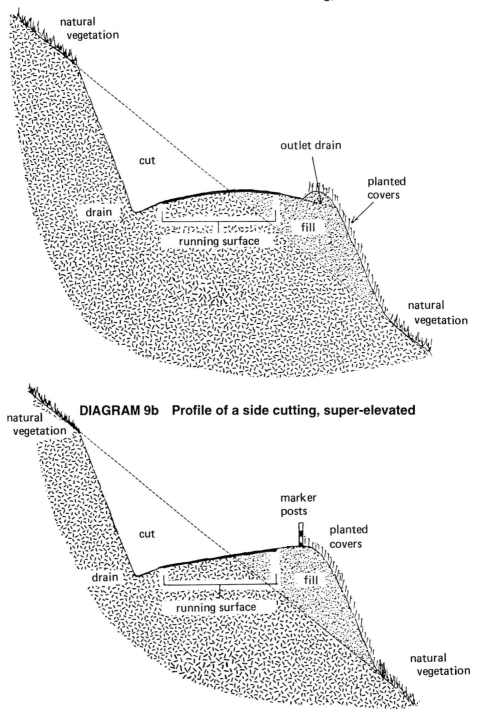

DIAGRAM 9b   Profile of a side cutting, super-elevated

by a grader, only the profile in Diagram 9b should be used. Even then it is suggested that marker posts are placed on the outside as a guide and warning to drivers, particularly at night.

The initial forming of a side cut is usually done with a lightweight dozer equipped with an angled A-blade. The steeper the hillside, the greater the advantage such a machine has over bigger and heavier units equipped with S-blades. For a harvesting road one such machine may be enough but on main roads heavier machines with S-blades and rippers should augment it. Where the hillside consists of soil suitable for fills or for gravelling, the use of a cut and carry team should be considered as quarrying out one's road can be a cheap way of getting two jobs done for the price of one.

Embankments

Diagrams 10a and 10b on page 88 show embankments on firm ground and 11a and 11b on page 89 embankments on wet ground. It is important, when an embankment on dry ground is built, that the whole width at the base is consolidated by the machines and not just the road width. Assume that the base width of the embankment will be the road formation width plus twice the height of the embankment and MARK IT OUT CLEARLY PRIOR TO THE MACHINES STARTING WORK. This will result in embankment sides at a slope of 1 in 1 which, if properly consolidated, will stand up unless the soil is very lacking in natural binder. In such a case an even wider base could be required to maintain a stable embankment. (e.g. the road formation width plus four times the height of the embankment to give a slope of 1 in 2). Having marked out the base of the embankment, drainage on either side, if required, can be dug and the spoil put into the embankment area. If the filling is from a nearby cutting, being pushed in directly by bulldozer, make sure that the machines follow through on their loads and, as they spread them, they run over what they spread to compact the soil. If the embankment is a long one, with material being brought in by a cut-and-fill team the same principles apply: use a lightweight dozer to spread the material and, if possible, a sheep's foot compactor working with the bulldozer continuously to pack down the soil layer by layer as it is added. It is a good idea to get the bulldozer to run backwards and forwards directly across the whole formation to ensure that the banks are firm right to the very edges before closing off at night.

When the embankment has been finished by the bulldozers it is possible to mark out the road edge with short posts for the grader to work to. These need be no more than $1/2$ metre outside the road as the grader can work to much finer tolerances than the bulldozers.

The same points with regard to kerbs or no kerbs apply on both sides of the embankment as apply to the outer side of a side-cut formation, except that it is impossible to comfort the driver by leaning him away from the drop over the side – the road must be cambered to drain the water off.

On wet ground where drainage is possible (Diagrams 11a and 11b), one method of doing the job is to cut drains a metre or two outside the intended road formation, throw all the spoil onto the formation and crudely camber it off to ensure that it is self draining, perhaps as much as a year or more before the road is to be built. Check a while after doing it, preferably just after heavy rain, that no water is lying unable to run off into the drains. Drain off any puddles found

DIAGRAM 10a   Profile of an embankment on hard ground, with earth kerbs

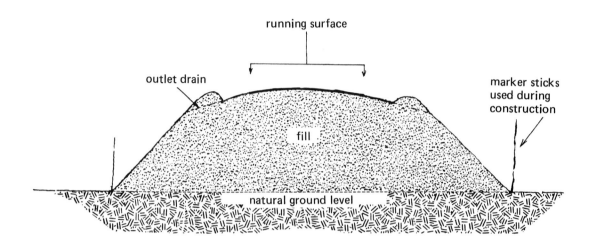

DIAGRAM 10b   Profile of an embankment on hard ground, open

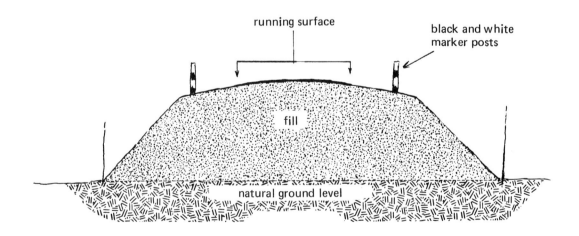

DIAGRAM 11a   Profile of an embankment on soft ground, with soil kerbs

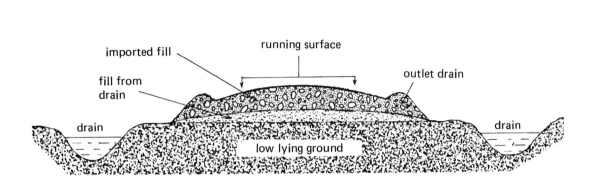

DIAGRAM 11b   Profile of an embankment on soft ground, open

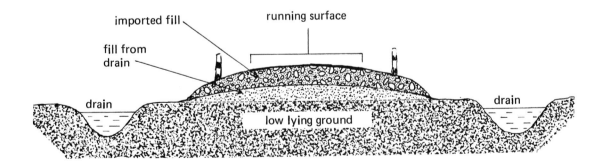

DIAGRAM 11c   Profile of a corduroyed embankment on wet ground

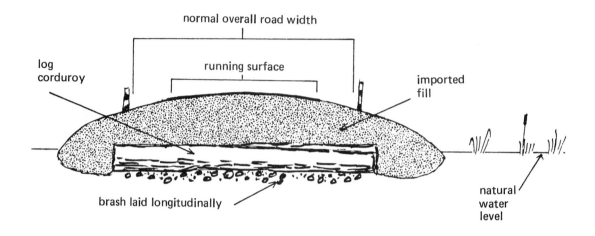

DIAGRAM 12   Profile of a cambered super-elevation for a curve

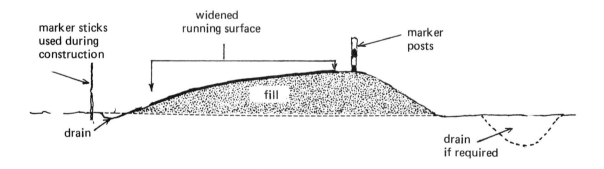

at the time by hand if necessary. When it has firmed up, the embankment so created can be shaped and then dressed with gravel to form the road or it can be blanketed with a thick layer of good soil before shaping and dressing.

If the ground is too sloppy to be worked by this method, and is undrainable, then it may be necessary to corduroy the road with logs first and to cover these with a thick blanket of soil. If this method is used (see Diagram 11c on page 90 and Picture 3.7 on page 74) the timber laid must be completely covered or below water level so that anaerobic conditions prevent the timber from rotting. If the ground is drained at a later date and aerobic conditions arise where the logs lay they will rot and the road will be affected by subsidence as a result. This is however, a practical way of making a road over a swamp. Again, mark out as above, cut the timber to be used for the corduroy at least a metre shorter either side than the base width (to ensure its complete burial away from the air) and lay the logs down butt to top alternately, except when a curve is encountered, when all butts should be on the outside of the corner. If the ground is very soft it may be necessary to lay brash down longtitudinally before the logs are laid. This would have to be done by hand. Logs anything up to a metre in diameter could be used in such circumstances but do be sure that the diameters of adjacent logs are fairly similar to provide an even bed (see again Picture 3.7). The ideal machine for this work is a tracked loader equipped with log forks but, if one is not available, a bulldozer can marshall the logs into place pressing each one down carefully with its blade as it goes in. Poles may have to be used as runners to enable the dozer to push new logs over those already laid. The dozer will help to level off the logs as it tracks over them. For a permanent road the earth fill over the logs should be not less than 60 centimetres thick spread as a single layer by bulldozer. The choice of soil is important – a well structured heavy red tropical clay is best. If possible top-off with rotten rock and then dress with gravel.

Curves and gradients

Gentle curves

The degree of super-elevation required on a corner is related to the sharpness of the corner and the speed at which the vehicles travel around it. The sharper the curve or the faster the speed, the greater the required degree of super-elevation. Because the rear wheels of normal vehicles track within the path of the front wheels on a corner, corners need to be wider than straight road and the sharper the corner the wider it needs to be. However, it is obvious that if the road is super-elevated to slope all one way, all the rain water from the upper half of the road must pass over the lower half in order to get into the side drains. This doubles the quantity of water draining over the lower half. If the road has to be widened still further then this aggravates the situation. Add to that the fact that the ability of flowing water to transport material increases by the sixth power of the speed of flow (see *River control* page 212) and there are good arguments for making curves as gentle as possible.

It will be observed that wide super-elevated corners on roads that are not heavily used, ripple very badly on the inner side of the corner. Heavily used roads where the super-elevation is inadequate will show a ridge of loose gravel piled high on the outer edge, whilst the inner side will be badly worn by scuffing tyres. A simple way of finding out whether the super-elevation is adequate, is to ride or observe a motorcycle taking the corner at what is considered to be a reasonable speed and see whether the motorcycle is perpendicular to the road surface or not. If the motorcycle leans in (relative to the road surface) on the corner, the super-elevation is inadequate, if it appears to lean out then the super-elevation is too great. One should aim to err on the side of having the super-elevation slightly low and then the tendency of the vehicles to throw gravel to the outer side will counter the effect of the rainwater washing the soil down to the inner side. Some sort of balance which will reduce the need for maintenance can thereby be achieved.

A profile for a gentle curve on flat land is shown in Diagram 12 on page 90. It will be noted that a camber is still present, superimposed on the super-elevation. This is not unreasonable as the vehicle travelling on the inside of the bend is negotiating a sharper curve than the vehicle on the outer side. It is as well to place the connecting drain a metre or two outside the actual road formation. Note also that the outer super-elevation is achieved without deepening the drain edge on the inside corner; earth has to be drawn along from the straights either side, or brought in for the purpose, otherwise water will lay on the inside of the corner which will then become soft and muddy.

The effects of gradient on a gentle curve

Where high-loaded vehicles are to use the road (log lorries or tippers stacked high with oil palm fruit for example), one should really consider the speed of the vehicles travelling up and down hills. A loaded lorry crawling up a steep hill will not need any super-elevation to ensure its stability, whereas going downhill fully loaded it may travel very fast and will need the effects of super-elevation in order for it to remain stable.

On the assumption that the lorries are travelling in one direction loaded and returning empty – a common enough circumstance – Diagram 13 on page 93 shows for the four possible combinations, the ideal super-elevations for a heavily loaded lorry travelling on the left hand side of the road. The "up left" situation is the one which provides the problems because it indicates the need for a "hollow" road which is unacceptable from the point of view of draining it. The other three situations are all acceptable from that viewpoint and, as the vehicle that is returning light is both more stable inherently, and able to climb hills at speed, the theoretical requirements are easily translated into practical super-elevations with or without cambers superimposed upon them.

If the "up left" situation is a real problem on particularly busy and dangerous hills, it may be necessary to separate the carriageways as shown in Diagram 14 on page 93, but note there must be sufficient width in each carriageway for vehicles to overtake one another so that if a vehicle breaks down it does not close the carriageway. Drainage between the carriageways is also necessary with relief culverts (shown dotted in Diagram 14) to take the water away underneath one or other carriageway. This is an expensive option but the alternatives are:

DIAGRAM 13   The effects of gradient upon curves: the "up left" problem

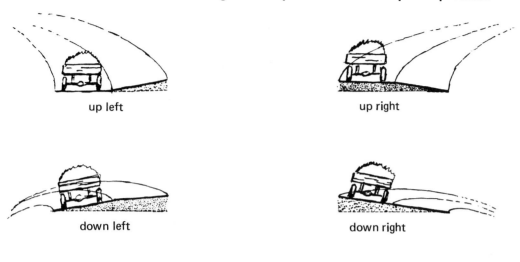

up left   up right

down left   down right

DIAGRAM 14   The split carriageway solution to the "up left" problem

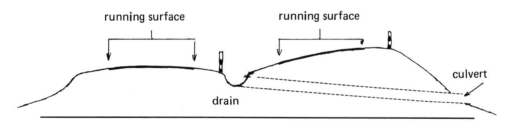

running surface   running surface

culvert

drain

- To super-elevate, which may give up-going lorries real problems in very wet or very dry conditions.
- To provide the road with a normal camber, which will encourage down-coming vehicles to cut the corner for fear of falling off the edge to their left and encourage up-going lorries to cling to the centre or right hand side of the road, in order to retain stability. Both are dangerous driving practices.

With problems like this there is plenty of incentive for keeping gradients down to acceptable levels especially in the "up left" situation (or its converse for those who drive on the right hand side of the road).

Gradients on corners and junctions

The sharper the corner, the more necessary it is to keep the gradient as low as possible (see *Working with the field survey team*, page 15). It should be a tenet of earth road design that steep gradients should be ascended or descended on straight sections of road, gentle gradients on either straight or gently curved sections, and that the road should as nearly as possible level off

DIAGRAM 15   Profile for a corner, single carriageway

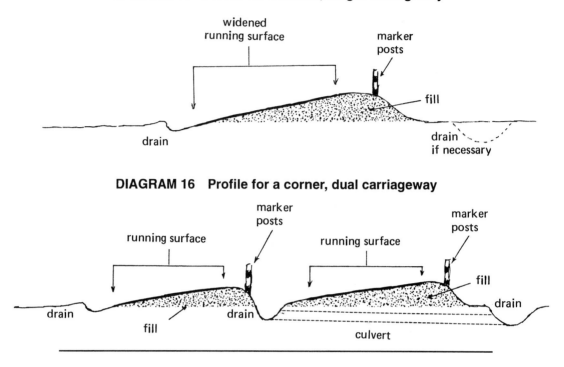

DIAGRAM 16   Profile for a corner, dual carriageway

some ten or more metres before entering a corner, remaining fairly level through the corner and for another ten metres or so beyond it. Level in this context means gradients of less than 1 in 30. Exactly the same consideration should apply to junctions.

Profiles for corners

The profiles for corners, being sharper than curves, will be super-elevated as shown in Diagram 15, above (compare with Diagram 12, on page 90) which shows a simple single carriageway. They will suffer the problems of erosion and wear noted in the first paragraph of the section on *Gentle curves* above, aggravated by the need for greater and greater width the sharper the corner becomes. It can pay, both as a means of reducing maintenance costs and increasing safety, to split the road into two carriageways as shown in Diagram 16, above. Since corners are by their nature short, this is not an unduly expensive exercise. It will require only one or perhaps two drainage pipes (shown dotted in Diagram 16) to remove rainwater from between the two lanes, and not a great deal of extra land.

Leading into and out of corners

Coming from a straight section of road into a corner involves a change of profile from an evenly balanced camber to a full super-elevation and then, as one emerges from the corner, reverting back to an evenly balanced camber again. However, the two should not be symmetrical: for the vehicle travelling on the outer side of the curve, there must be a difference. Imagine first the vehicle on the inside of the corner: it is leaning in the appropriate direction before it enters the curve, continues to lean the correct way during the curve and emerges from the curve onto the camber still leaning the same way. On the other hand, the vehicle on the outside of the corner is leaning outwards relative to the corner before it enters it. The road should change profile to provide for the vehicle to lean inwards just as the vehicle starts to corner, should continue that way through the corner and revert to leaning the vehicle the other way again as required for the normal camber of the road ONLY WHEN THE VEHICLE HAS FULLY COMPLETED CORNERING (see Diagram 18 on page 98). Failure to carry the super-elevation into the first few metres of the straight section of the road, is a prime cause of accidents in which vehicles fail to complete a corner by rolling or sliding off the road at the end of the corner. This is because a driver will enter a corner and naturally regulate his speed as he goes around it to one at which the vehicle corners comfortably. This speed can be too fast for safety if the adverse camber is resumed on his side of the road before the vehicle has fully completed the curve. If he attempts to brake on what is likely to be a loose surface he will only worsen the situation. Starting the super-elevation early, before the vehicle has left the straight and started entering the corner, is also bad practice, though not as bad as ending the super-elevation too soon. It will either cause the vehicle to wander towards the centre of the road and encourage cutting of the corner or will encourage the driver to enter the corner at an unwisely high speed, both of which can contribute to the risk of accidents.

When forming super-elevated corners with a bulldozer it is more efficient to work with an S-blade and hydraulic tilt than with an outwardly angled A-blade (see Diagrams 17a and 17b on pages 96 & 97). On entering the corner as shown in Diagram 17a the bulldozer should tilt the blade right corner down, cut where the arrow "a" is, lift out through arrow "b" and fill through arrow "c" to point 1. The operator should then back up, change angle a little to the right and repeat the process through the three arrows to point 2, then to point 3 and so on sequentially until the corner is finished. By the time the first two cuts are done, the angle of the super-elevation will have been set and because the bulldozer will be itself tilted inwards, work on subsequent steps will be done with the blade in the "level" position relative to the bulldozer. It will be noted that by working to the right, the super-elevation will be carried through well into the beginning of the straight sector of the road as shown in Diagram 18 (see *Leading into and out of corners*, page 94 above). This is not to imply that working in the opposite direction is unacceptable but to point out that working to the right does save time and work if there are no other problems to consider (vice versa for countries where driving on the right is the rule). The advantage of this tangential method of working for the machine is that full power is available on both tracks as the machine is running straight whilst cutting. Also that the C-frame or push arms are not rubbing against the edge of the road as is the case when doing the same job with an A-blade outwardly angled and the bulldozer following round the corner (see Diagram 17b). Operators should be taught to use the method shown in Diagram 17a.

DIAGRAM 17a　Forming super-elevated corners: correct method

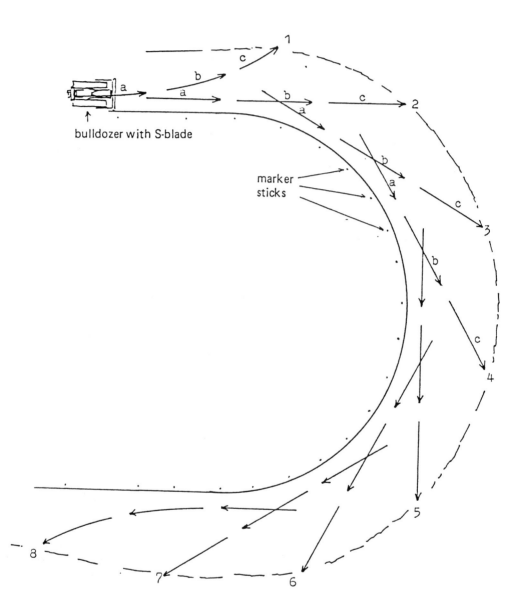

bulldozer with S-blade

marker
sticks

DIAGRAM 17b   Forming super-elevated corners: incorrect method

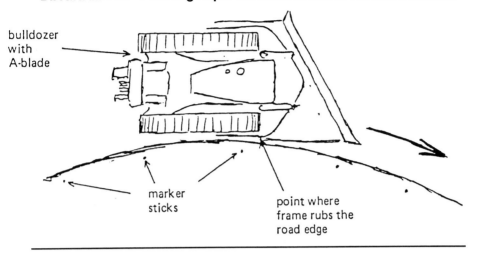

bulldozer with A-blade

marker sticks

point where frame rubs the road edge

Corners on side cuts

Two situations can exist: where the corner turns into the face of the hill (Diagram 19 on page 99) and where the corner turns away from the face of the hill (Diagram 20 on page 99). In both cases rainwater, mud and rock can fall onto the road from the hillface. In the situation in Diagram 19, whilst landslides may block the road for traffic, it takes very considerable slides to overtop the road edge and send subsequent flows down the face below. It is also usual when the corner turns into the face of the hill that the hill will have been substantially cut away and the earth from the cutting pushed into the next gulley to form the road where the corner turns away from the face. As a result, the corner turning into the hill will be largely founded on undisturbed ground and, even if the road is overtopped, the damage done by the water flowing onto the ridge below will be minimal. The Diagram 19 situation is therefore usually inherently more stable than the Diagram 20 situation. Where the corner turns into the hill, adequate width should be left for the drain on the inside of the corner if possible – and one has to recognise that very often it is just not economically possible to get as much space as one would like.

The corner turning away from the face of the hill is often the place where a culvert in a stream bed will be sited, as in Diagram 21 on page 100. Even if this is not the case, there will still be a tendency for a large proportion of the road to be founded on fill. Assuming that what lies beneath is not precipitous there is likely to be more room available for roadside drainage and, at first thought, to receive any landslides that may occur. However, since the road will be super-elevated away from the face of the hill, even a small landslide, enough to block the drain and verge, will send water cascading across and over the edge of the road where it can be expected to erode the fill upon which the road is based. Then again (see Diagram 21) landslides are much more likely to occur where a substantial depth of soil has been cut than in places where little or none has been cut out of the side of the hill. However, in the writer's experience, the place where most troubles are likely to occur is at the transition between the two corners

DIAGRAM 18   Leading into and out of corners, from camber to super-elevation to camber

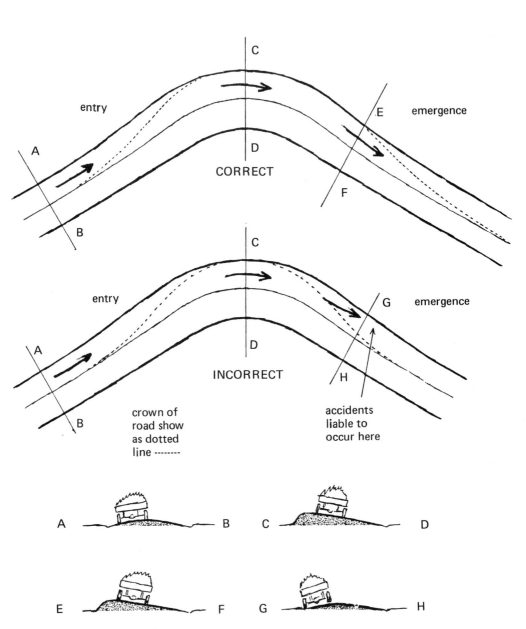

entry

A

CORRECT

B

C

D

E   emergence

F

entry

A

INCORRECT

B

crown of
road show
as dotted
line --------

C

D

G   emergence

H

accidents
liable to
occur here

A B   C D

E F   G H

DIAGRAM 19   Corner on a side cut: turning into the face of the hill

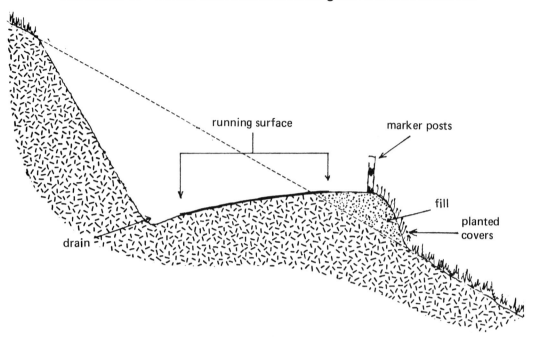

DIAGRAM 20   Corner on a side cut: turning away from the hill

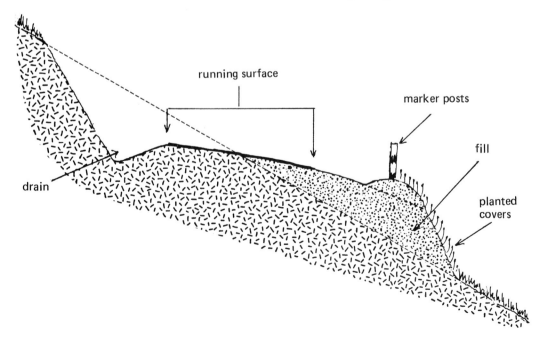

DIAGRAM 21   Corners on side cuts

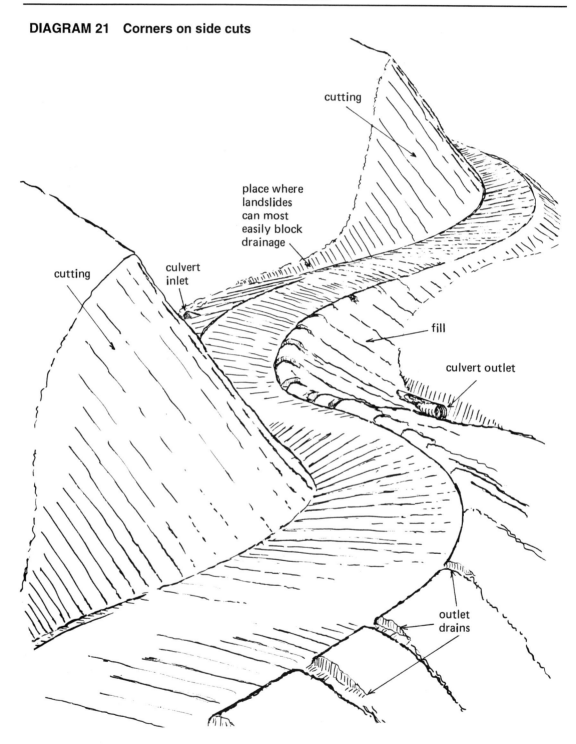

cutting

place where
landslides
can most
easily block
drainage

cutting

culvert
inlet

fill

culvert outlet

outlet
drains

(refer to Diagram 21 again). There the natural uncut ground tends to be at its steepest and the room to spare, since one is at that point channelling the water from the cutting to the culvert, the least and so most easily blocked. A small slide there can divert the water across the road to drain out over the fill where the potential for damage is considerable.

This raises the question – when transiting across a dissected hill face on a roughly level trace – of where one should aim to place the culverts. If there are hard, perhaps rocky stream beds, not liable to erode very readily, then the culverts are best placed there. They will receive both the water from the stream (which may be dry between rainfalls), and the road, and therefore one should design the road to rise between stream beds so that drainage falls naturally from each deep side cutting towards each stream bed. This brings with it the risk that if the culvert gets blocked, the gulley behind it will fill until it overflows the road where the culvert is sited and will then be well able to wash out the road, and perhaps the culvert with it. Such a risk may be deemed acceptable if the road maintenance is good and one is prepared to put in a large enough culvert to minimise the risk. If the situation is one in which the extra water draining in from the roads is likely to cause unacceptable additional erosion to the stream beds, then there is an argument for:

- Placing culverts in the streambeds to receive the water from them.
- Making the stream culvert sites the highest points along the road.
- Letting the road fall and rise between streams.
- Putting in culverts to take off the roadside drainage water at the lowest points on the road. They would then release their water onto the harder uneroded ground of the ridges between the streams and below the road.

This method requires more culverts but the discharge from each will be much less and so they can be of smaller diameter. It will also mean that the risks of landslides diverting the flow of water across the road where it consists of fill will be averted – overtopping will only occur where damage will be minimal. It can also mean that the road will be lengthened by a small amount as the trace reaches down the ridges and up the gulleys, whereas a trace rising over the ridges and falling into the gulleys tends to shorten the road.

Note: a perfectly level road on a side cut is always a nuisance to maintain because the slightest land slip will block the drainage and cause water to lie there for a long time. This is because there is no gradient to give the water enough flow to start to wash away the loose soil of the slip. It is always better to let the road rise and fall at gradients of around 1 in 20 or 1 in 30 between culvert sites.

When the road is climbing a hill at a steady gradient and the hillside is crossed by streams, one is faced with the situation in which the blockage of any one culvert will lead to overflowing down the side of the road to the next culvert. This in itself can do a great deal of damage to the road (see Pictures 10.3 to 10.7 on pages 250 to 252) and, if the next culvert down is unable to cope with the extra flow, the problem may extend until the bottom of the hill is reached. This can be avoided by maintaining the rising gradient up to the culvert site and over the culvert but dipping on the uphill side of the culvert before resuming the gradient again (see Diagram 22 on page 103). This will provide a low point which will be overtopped by flood water if the culvert is blocked. If this gets washed out it will not affect the culvert or the road downhill from it and minimal damage will be done, both to the road and the land to the side of it. Such damage to the road will be easy to repair afterwards.

Note too in Diagram 22 the dotted line indicating that the gradient of the roadside drain should be maintained. This is done in order to prevent the drain becoming silted up as would happen if the gradient was to be reduced. It is important that when designing a road the gradients of roadside drains should have a uniform or a convex longitudinal profile (see Diagram 23a on page 104) and not a concave profile (Diagram 23b). Silting will occur on a concave profile and can spread into the road or even overtop it with subsequent damage. If this situation cannot be avoided, then it will be necessary to place a supplementary culvert at the point a little way below where the gradient becomes inadequate to maintain flow speeds (Diagram 23c). Where this point will occur will depend on the nature of the soil being worked and is best learnt by observation on site.

Hairpin bends

In order to gain height on a side cut it may be necessary to construct hairpin bends. The need to select sites for these with care has already been mentioned (see *Working with the field survey team*, pages 15 and 16). Assuming that a suitable site is available to be marked out, one of the principal problems will be that of drainage resulting from the change of gradient upon leading into the bend from the upper side (see Picture 3.17 on page 79 and Diagram 24 on page 105). There are two options:

- To dig a drain around the bend on the outside of the bend until the water can be thrown off (dotted line in Diagram 24a, drain shown on 24b).
- To put one culvert under the formation at the upper end and another under the lower end of the corner (Diagrams 24a and 24b) connected by a short drain which will also receive water from the inside edge of the bend itself and discharge it down the hillside.

The advantage of the first method lies in saving the cost of the culverts. It will also result in throwing the water off the end of the formation which could be an advantage, depending on where the water will flow from there. It is necessary, if there are several hairpin bends in the road stacked above one another, to try and avoid throwing water off one bend only to have it flow onto the road again lower down. The disadvantage of draining around the bend is that the side cutting above the bend to make room for the drain has to be deeper and wider, bringing with it a greater risk of landslips and thus a greater risk that water blocked by a landslip will overflow and damage the formation.

The advantages of the second method lie in saving the job of digging a deep runoff drain at a place where it may be very difficult to do, and of providing an immediate runoff for the water from the bend itself through the lower culvert (though there is no reason why a little culvert should not be placed in the lower position if the first option is taken). In Picture 3.17, the first

DIAGRAM 22   Provision of flood relief overflow alongside culvert

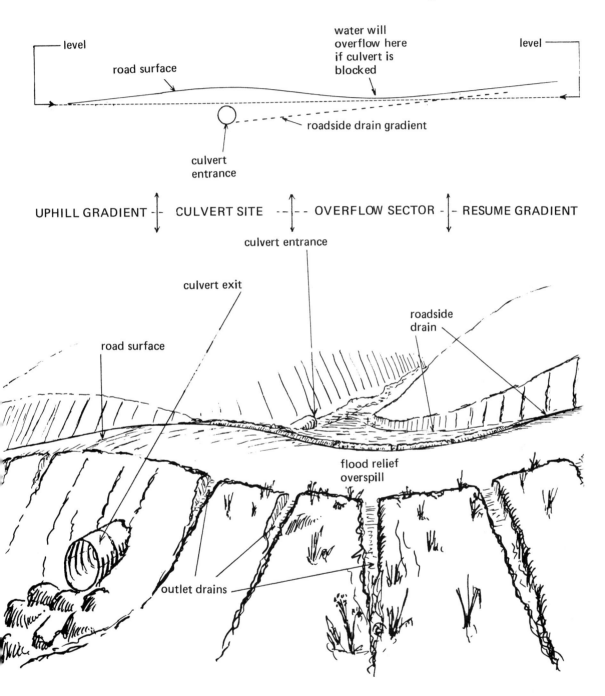

level

road surface

water will
overflow here
if culvert is
blocked

level

roadside drain gradient

culvert
entrance

UPHILL GRADIENT -|- CULVERT SITE -- -|-- OVERFLOW SECTOR -|- RESUME GRADIENT

culvert entrance

culvert exit

roadside
drain

road surface

flood relief
overspill

outlet drains

DIAGRAM 23a   Drainage: longitudinal profiles

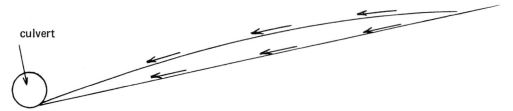

culvert

Because water flow speeds in roadside drains increase with either convex or uniform gradient profiles, no silting will occur before the water enters the culvert

DIAGRAM 23b

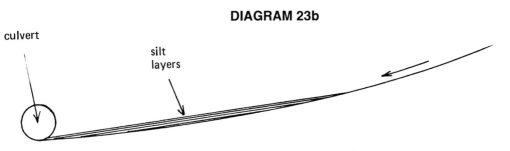

culvert

silt
layers

Because water flow speeds in roadside drains will fall on a concave gradient, silting will occur as the water slows down and will back up the drain until a gradient at which the water can continue to carry the silt is achieved At this point matters will stabilise unless the water over tops the banks or road first

DIAGRAM 23c

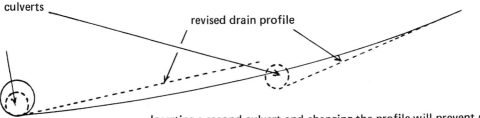

culverts

revised drain profile

Inserting a second culvert and changing the profile will prevent silting

DIAGRAM 24   The drainage of hairpin bends: alternatives of draining around the bend or culverting across the bend

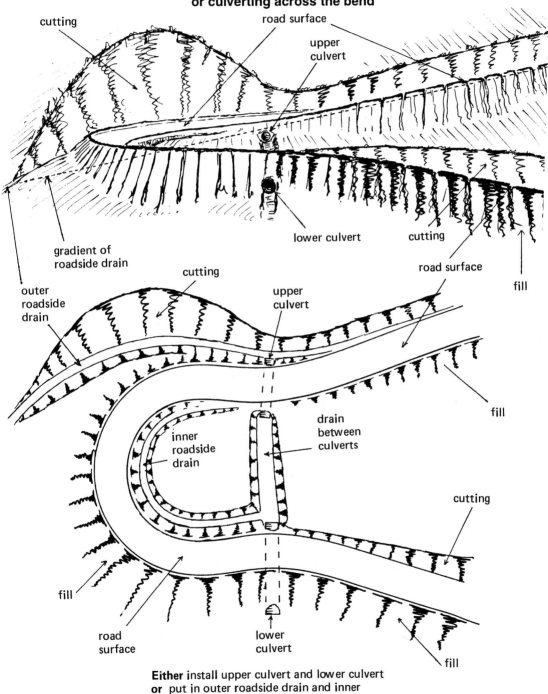

Either install upper culvert and lower culvert
or put in outer roadside drain and inner
roadside drain and lower culvert

option of draining round the outside of the curve was selected. A blue marked stick is used as the focus of the radius of the bend and unpainted sticks mark the inner radius, the lead in and lead out for the machines to work to. The gradient beyond the tree to the left falls away at 1 in 15 and the gradient from just in front of the vehicle rises to 1 in 15. The corner of the hairpin bend falls a little less than one metre through the entire inner radius of five metres. The road in the picture was built across a 350 metre high hill in Irian Jaya, Indonesia, to give access between two large areas of plantable land and involved the construction of about ten hairpin bends. No gradient steeper than 1 in 15 was allowed.

Junction layouts

Any junctions of the estate road system with the public road system should be designed in co-operation with the public road engineers concerned. They may require that several tens of metres of the project road adjoining their road is sealed, to reduce the extent to which vehicles leaving the project carry mud and grit onto their road's surface. If any entrance gate or control point is to be installed by the project, then there must be adequate room between the gate and the edge of the public road for the largest vehicle likely to enter the estate to stop, whilst the gate is opened or closed, without either obstructing the public road or blocking lines of sight essential for the safety of other road users. The entrance thus formed must also be wide enough for two such vehicles to use it simultaneously, one entering and one leaving, without the need for one to wait on the public road for the other to clear the junction.

Since junctions are sites at which traffic flows will converge and diverge and cross one another, they are focal points for accidents. Attention should therefore be given to ways and means of reducing the risks involved. Of prime importance is the provision of adequate visibility, both on the junction and for some distance either side of it. Junctions, like sharp corners, should be on the level as far as possible. Side roads should level off before joining main roads or should rise gently into a main road – there should be no risk of an out-of-control vehicle being able to run downhill onto a main road carriageway in front of other fast moving vehicles. A design conforming to this limitation is not only safer but will prevent rainwater from an ill-maintained side road from washing onto the main road.

Junctions on the inside of a bend (see Diagram 25a on page 107) are very dangerous unless clear visibility is assured over the whole of the shaded area indicated in the diagram and there is ample room between the points marked A and B for fast moving vehicles to stop without any risk of collision with a slow moving vehicle emerging from the side road. It is better in such a situation to divert the side road to come out and join the main road a little way along the straights at either end of the curve in the main road (Diagram 25b). There adequate visibility can be provided much more easily without sacrificing so much plantable land or having to maintain short cropped vegetation.

Another junction layout that can be dangerous is the "Y" junction (see Diagram 26a on page 108) where two main roads merge at a narrow angle. The problem lies with traffic travelling from B to C, which has to cross traffic moving from C to A, and may not be able to see traffic coming from A to C, because the driver will be in the right-hand side of the cab and his load

DIAGRAM 25a Layout of a junction on the inside of a bend

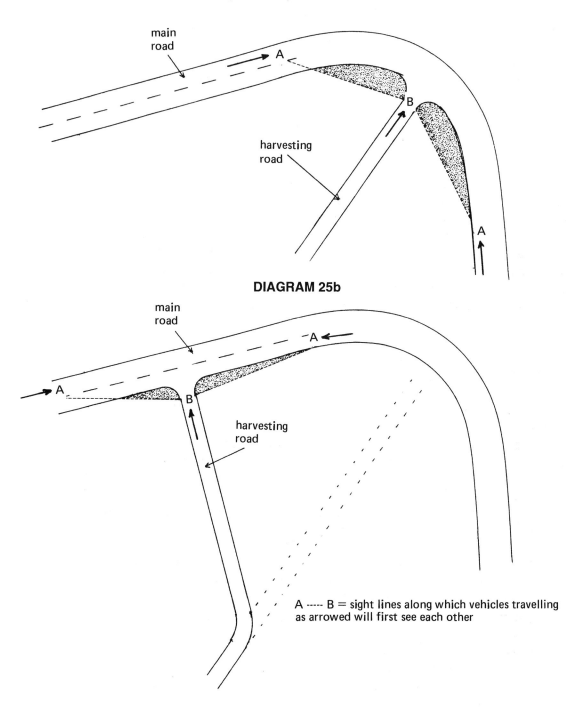

DIAGRAM 25b

A ----- B = sight lines along which vehicles travelling as arrowed will first see each other

DIAGRAM 26a "Y" junction layouts

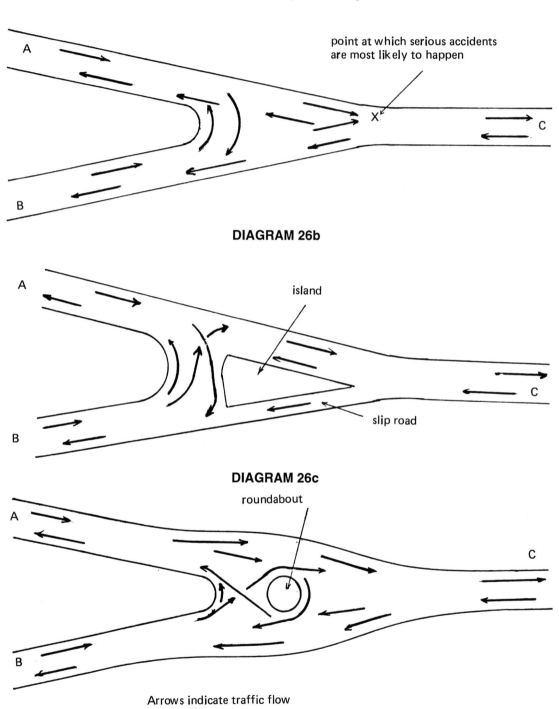

Arrows indicate traffic flow

will obstruct his view (vice versa for those who drive on the right). The danger point is at "X". One solution is to make A to X the minor road and B to C the major road. Another solution is to make traffic from B meet the A to C road at right-angles (see Diagram 26b) which will force vehicles to slow down almost to a stop for what becomes virtually a normal "T" junction. Traffic from C to B then uses a slip road and gives way to the easily seen traffic from A to B (which by the very nature of a converging "Y" junction, is likely to be negligible – most traffic will have found an earlier cross route to get from A to B). Only a small island has to be maintained to allow full visibility for the traffic using the slip road and no great area of land is sacrificed. The other solution is to make a roundabout (see Diagram 26c), which does not have to be very large, but the roads should be wide enough to enable fast traffic from A to C and from C to B to slip by without too much interference from traffic having to go round the roundabout. A give-way-to-vehicles-on-the-right rule then has to be applied.

Roundabouts are more difficult to maintain with a grader than normal junctions unless they are very large. They also bring drainage difficulties (unless drainage can be arranged underneath the road from the centre) because logically the central ring of road surface should be super-elevated inwards. Even with proper super-elevation, wear on the road surface at a roundabout is extraordinarily heavy and so it is always a nuisance to upkeep. A clue to the cause of this lies in the fact that in the roundabout solution both the main traffic streams from C to A and from B to C have to turn round the roundabout. In the junction solution (Diagram 26b) only the B to C main traffic stream has its smooth progress affected. For this reason, and because it will be easier for the grader operator to work on, the junction solution in Diagram 26b will be cheaper to upkeep. Drainage from the island can be effected under the slip road if possible, which will mean using only a half-road-width long culvert; if in the other direction a full road-width long culvert will be required. The junction and island solution shown in Diagram 26b is, on balance, the preferred answer to the problem.

If a roundabout is deemed necessary then it is better to lay it out to minimise scrubbing and reduce the risk of accidents, bearing in mind that a loose surface can be a slippery one. Bear in mind also that the radius of the roundabout, and the lead-ins and lead-outs, must be adequate to enable both the grader to maintain, and the biggest vehicles using the road to negotiate, the system in comfort.

The normal roundabout layout is shown in Diagram 27a on page 110. "Give Way" signs will normally be placed to the left side of the road for convenience and to avoid drivers missing the sign when facing oncoming headlights at night from across the roundabout. Vehicles entering will give way to traffic on the right and initially turn left. If exiting first left, the driver will probably straighten-up, cutting across the carriageway to some extent, and then turn left again. If using any other exit he will first turn left to enter the roundabout, turn right to negotiate it and then turn left to exit. He will probably accelerate as he leaves and also probably swing across towards, even into, the lane by which oncoming vehicles enter the roundabout. This potential accident spot is marked *X* in the diagram.

The layout suggested in Diagram 27b avoids the risk of accidents present in the normal layout. Vehicles entering will have to slow down, may have a "Give Way" sign immediately in front of them which they cannot fail to see, and turn left. A vehicle leaving first left will then

DIAGRAM 27a   Normal roundabout layout

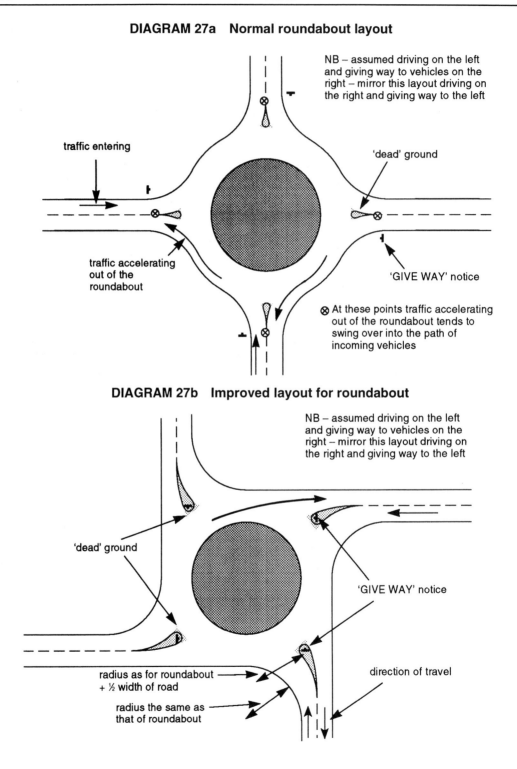

NB – assumed driving on the left
and giving way to vehicles on the
right – mirror this layout driving on
the right and giving way to the left

traffic entering

'dead' ground

traffic accelerating
out of the
roundabout

'GIVE WAY' notice

⊗ At these points traffic accelerating
out of the roundabout tends to
swing over into the path of
incoming vehicles

DIAGRAM 27b   Improved layout for roundabout

NB – assumed driving on the left
and giving way to vehicles on the
right – mirror this layout driving on
the right and giving way to the left

'dead' ground

'GIVE WAY' notice

radius as for roundabout
+ ½ width of road

radius the same as
that of roundabout

direction of travel

drive straight out of the roundabout without having to turn further in any direction. A vehicle leaving by any other exit will turn left on entering, turn right to negotiate the roundabout and straighten-up to exit. Note that any vehicle exiting has no reason to swing towards or into the path of oncoming vehicles entering the system. Furthermore, the driver has only to turn the steering wheel twice, left then right, not three times: left, right and then left as in the normal layout. This will reduce wear on the road and the vehicle, and make driving easier and safer.

Some tips and ideas:

- When designing junctions always remember to check the minimum acceptable radius required by the vehicles that will use the road, AND the minimum radius that the grader can conveniently operate round. Make the largest of those figures into the minimum radius to be used – often it will be found that it is the grader that will place the limit on the minimum radius that it is practical to use. Failure to consider the requirements of the grader is a common cause for the unsatisfactory upkeep of junctions.

- Ensure that the centre of the road surface of the junction is the highest point so that water will always drain naturally away from it into the roadside and outlet drains.

- Always look critically at drainage. If possible raise the whole junction just enough to ensure that rainwater runs back down the roadside drains of each contributing road away from the junction. If that is not possible, make sure that adequate culverts are placed and that the drains leading into the culverts are deep enough to draw all the water off. Make sure that the culverts are buried deep enough never to be caught by the corner of the grader's blade or the tips of its rippers.

- The provision of visibility by leaving land unplanted for junctions within the harvesting road where vehicle speeds are low is not of critical importance. Where harvesting roads join main roads it is important, and on main road junctions it is vital, that visibility should be enough to enable the fast-moving vehicles on the main road to stop before colliding with any vehicles or pedestrians emerging from the plantations.

- Just as junctions can be obscured by vegetation, so corners can be hidden by hilltops (see Pictures 4.1 & 4.2 on page 112). On fast roads, changes of direction should only be made where drivers can see and assess them long before they begin to negotiate them. Corners sited on humps in a road's vertical alignment are inherently dangerous, as they catch unawares the driver who is unfamiliar with the road. Conversely, changes of direction situated where the vertical alignment is flat or hollow are safe.

Picture Group 4:   Alignment for safety/drainage problems

4.1/4.2   Changes of direction over a hilltop can be very deceptive. The road in the top picture does not go through the gap between the trees behind the man standing in the middle of the picture, but turns sharply left – a vehicle can be seen parked on the road in front of the tree at the left. In the lower picture the change of direction in the road is clearly visible long before the driver reaches it. This is much safer.

4.3/4.4   Junction layout. Picture 4.3 (top) shows how *not* to do it – meeting a public main road at a steep incline is dangerous. Picture 4.4 is a safe layout.

4.5   Care should be taken to avoid alignments which it is difficult to drain. For this case the only way to drain the road is by providing soak-away pits on either side, using the soil so gained to raise the road formations.

4.6/4.7   These two pictures show catastrophic erosion at gentle gradients (less than 1 in 20) on loess type soils. Normal road formation and drainage practices had been used, but proved to be totally inadequate in the face of a complete lack of clay-bound gravels with which to dress the running surface. Rainfall in the area is not particularly high, but tends to come in short heavy storms with disastrous results. The only way to control such a situation is with numerous soak-away pits, adequate to contain the runoff and prevent the water flowing in great volumes at any speed.

Railway crossings

Always cross a railway on level ground or, preferably, by dropping slightly onto a track and rising off it after crossing. If the track is higher than the road, and the road leads up and over it, it is very difficult for a grader driver to avoid, at some time or other, snagging the rails accidentally with his blade. This could damage the rail and lead to a train being subsequently derailed. Take care to ensure that the road drainage system does not adversely affect the railway's formation. After maintenance, the line should always be cleared so that nothing whatsoever remains on the rails that could endanger a passing train.

Bear in mind the "blind spot" on a lorry. The road should always be laid out so that it approaches a rail line at right angles on the blind or nearside. On the offside, oblique approaches are permissible. Ensure that visibility is adequate: trains cannot stop in a short distance even in an emergency.

Turning points

There are likely to be some spur roads in the harvesting road system. They must terminate with turning points which can be easily used by the harvesting vehicles – and negotiated by the maintenance grader too. A circle of adequate radius to be practical is not always the most economical answer and will require culverting to drain the centre. It is likely that a "Y"-shaped turning point with the branches both comfortably longer than the longest vehicle to use them and, with the road, meeting at 120° spacings, will be less demanding of land and easier to drain. However, a "Y" turning point will require the drivers of the vehicles to be able to reverse. This is easy with simple lorries, more demanding of skill with articulated lorries and tractors with two-wheeled trailers, but not suitable at all for four-wheel trailers or trailer trains. The road team management should try out for themselves whichever type is proposed, with the vehicles to be used on it, before finalising the design for general use on the project.

Drainage

The importance of effective drainage has been emphasised consistently in the notes above. The design objectives should be to get good drainage, without undue erosion, in a form that can easily be inspected and maintained. The formation of the roadside drain by extending the camber to make one integral shape enables the grader to maintain both the road surface and the drain simultaneously. This is good practice and is the preferred method wherever possible. By extension it is desirable to make the run-off drains from the roadside in such a way that they too can be maintained by the grader. When the road is on a rounded ridge top, this is easy to do by getting a bulldozer to diverge out and away from the main formation, cutting a "V"-shaped drain to lead water off to a point at which it will flow away naturally (see Diagram 28 on page 117). A grader can then follow through during construction without any supplementary hand work, provided the grader is equipped with a bulldozer blade, and on the assumption that the runoff has not been planted over. If harvesting road runoffs are planted over, upkeep by grader will be limited to the main roads where, in the interests of visibility, there will be some space between the plantings and the road edge. Harvesting road runoffs can be cut initially this way prior to planting and subsequently upkept by hand; it is only necessary to see that during planting no plants are grown right in the centre of the runoff drain and are placed no nearer than a metre either side of it.

Siting of the runoff drains should be affected by such considerations as:

- Changes in gradient, for example from a steeper gradient to a gentler gradient, to prevent roadside silting.
- Changes in formation, for example from a cutting to an embankment (see Diagram 29 on page 118).
- Whether a suitable discharge point can be found which is both right topographically and able to receive the water discharged without undue adverse effects (borrow pits made during the construction of the road formation can be very useful in receiving drain water and will act as silt pits at the same time).
- The distance from the last runoff in relation to the amount of water likely to be carried by the roadside drain during a heavy rainstorm – the object being to prevent the quantity becoming sufficient to scour the side of the road or cause erosion when discharged from it.

Like so many aspects of road making one has therefore to be constantly looking for opportunities: in this case where the runoffs can be made satisfactorily, because it is better to have many runoffs, each discharging a little water, than a few, each discharging a lot.

Where a road has to cross flat land it is well to try to integrate the road drains into the main drainage system for the area from the beginning. Wet drains should always be cut from the outfall upwards. The sort of road profiles shown in Diagrams 11a and 11b on page 89 are likely to apply. If drains parallel to the road are necessary they should be placed far enough from the road formation to avoid any risk of erosion in their banks undermining the road formation. If the road is likely to be up-graded and widened in the future, give allowance for that too.

Bridge spans should be well clear of the flood levels likely to be reached by any rivers that they cross. This can mean the construction of ramps up to their abutments. Where ramps have to be built remember to allow additional width between the roadside drains to provide for the sides of the ramps to be sloped at an angle that will hold satisfactorily (see Diagram 30 on page 119). Drains in flat land will be preferably dug with a backhoe of one sort or another. It is important that the sides of the drains should be laid well back to prevent them from falling in. The writer has a preference for the wide open "U"-shaped profile (see Pictures 3.9 & 3.10 on page 75) which is easily cut by the larger backhoes working from the road formation. If the ground is boggy a tracked machine is preferable and flotation can be improved by supplying the operator with twenty or thirty stout poles about half as long again as the tracks are wide, to be used under the tracks as a mattress. They can be picked up from behind and placed again at the front of the machine as work progresses. In such conditions it may be impossible to make a nicely shaped drain at the first attempt and it may be necessary to cut a shallow drain first, to establish some relief from the water, and return again some months later, preferably during the dry season if there is one, to redig a proper drain when some degree of drying out and consolidation has been achieved.

DIAGRAM 28   Roadside runoff drain cut by machine and maintained by grader: level or gently sloping ground

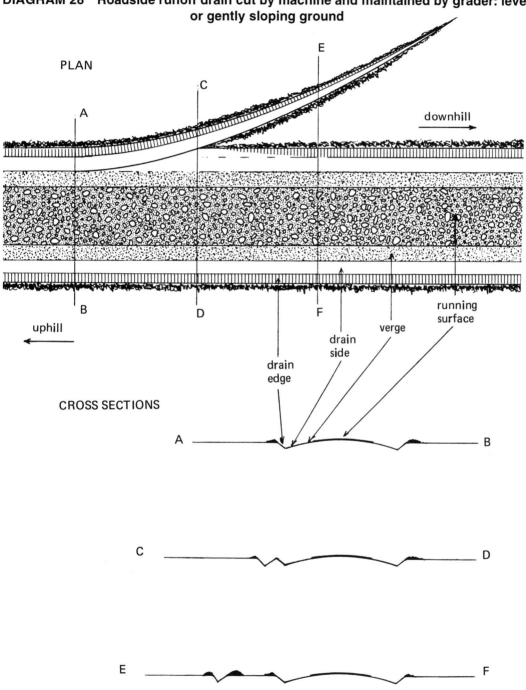

PLAN

E

C

A

downhill

B

D

F

uphill

drain
edge

drain
side

verge

running
surface

CROSS SECTIONS

A —————————————————— B

C —————————————————— D

E —————————————————— F

DIAGRAM 29   Roadside runoff drain cut by machine and maintained by grader: transition from cutting to embankment

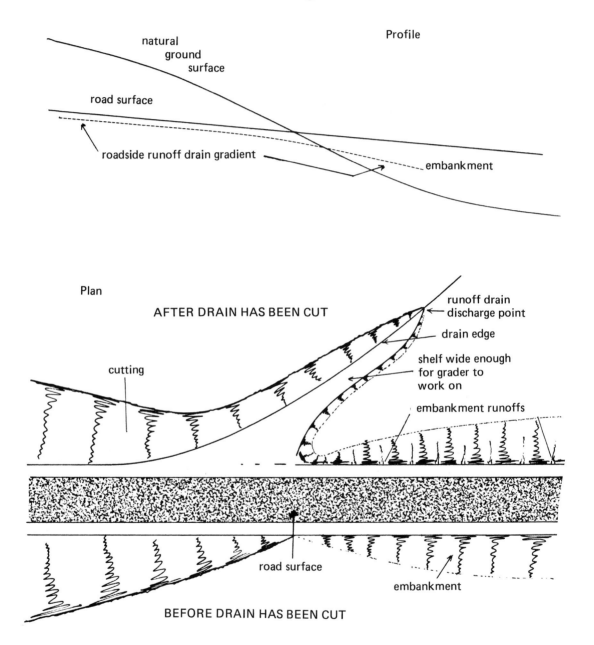

DIAGRAM 30   Ramping up to a bridge to keep the span clear of floods

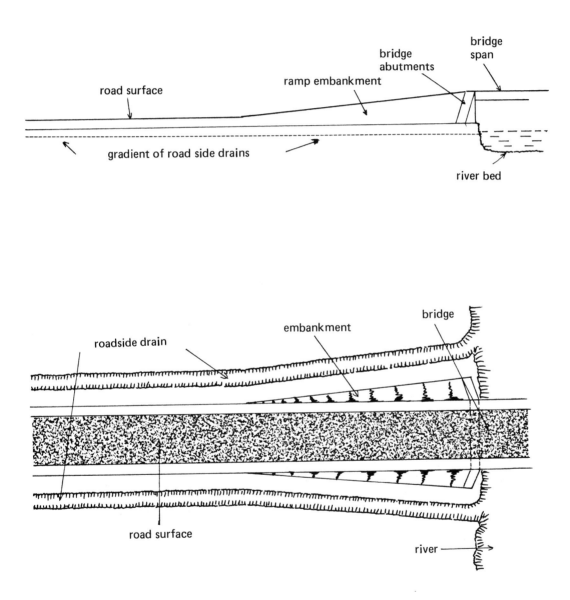

In rolling country where many cuttings and embankments have to be made when one builds a road across the grain of the land, long cuttings will gather a lot of water and there is no way of getting rid of it until the end of the cutting has been reached. It may be necessary to widen the total formation towards the cutting outlet in order that side drains of adequate capacity can be provided. When making a runoff drain as shown in Diagram 28 and 29, it is essential to cut into hard ground and ensure that the discharge point is well away from the embankment and that the water flowing from the discharge is not able to flow back towards the embankment and erode it. The gradient of the runoff must be commensurate with the gradient of the roadside drain feeding into it, neither too steep so that water can cut back towards the road by eroding the floor of the runoff (see Pictures 10.8 & 10.10 on pages 252 & 253) nor too gentle with the result that it will silt up readily. If it is impossible to avoid the runoff being too steep, the simplest solution is to let the drain erode to some extent and then fill it with large stone, too coarse and heavy for the current to move, and allow the stone to dissipate the power of the flow within the eroded channel. If no such stone is available then cement channels may be necessary but they are an expensive option initially and may be a source of trouble during subsequent maintenance.

Runoff drains discharged into plantings are usually no great problem – the plantations and their cover crops will adequately hold the soil. Where there is risk of erosion resulting from the release of the water, the planting of *Desmodium ovalifolium* which has the advantages of being both shade tolerant and easy to keep under control, is the writer's first preference as a counter measure. If not available, creeping covers with or without bush covers mixed in, or grasses, planted specifically as an anti-erosion measure, are recommended.

Coping with very-readily eroded soils

Where erosion is a very serious problem the whole emphasis must be on minimising water flow, both in terms of speed and of volume. Gradients should be the gentlest possible. Soils as susceptible as this, will be almost entirely lacking in binding clays and erosion can become serious at inclines of no more than 1 in 20 (see Pictures 4.4 & 4.5 on page 114).

The best method of coping with this problem will be to make borrow pits, approximately at right angles to the road, with which to provide soil to make the formation entirely proud of the surrounding land surface (see Diagram 31 on page 121). These borrow pits are intended to become sinks for drainage. They therefore need to be deep and big enough to contain all the rainwater that will be fed into them from the road. Their frequency will limit water movement and scouring along the edge of the road. They will need regular clearing out during maintenance. It is essential that they not be allowed to overflow or the water from one may flow down to the next one, probably causing that to overflow and so on. The resulting cascade from pit to pit will then cause just the kind of serious scouring and erosion that the system is designed to prevent.

The pits will retain such fines as are naturally present in the soil that gets washed into them. These, along with the silt, should be picked up by the front-end loader, put back onto the road and graded *uphill* to remake the road surface during maintenance. Any sticks or other detritus should be thrown off by labourers working with the grader. If the pits are not cleared regularly,

DIAGRAM 31   Use of borrow pits to facilitate soak-away drainage on level ground

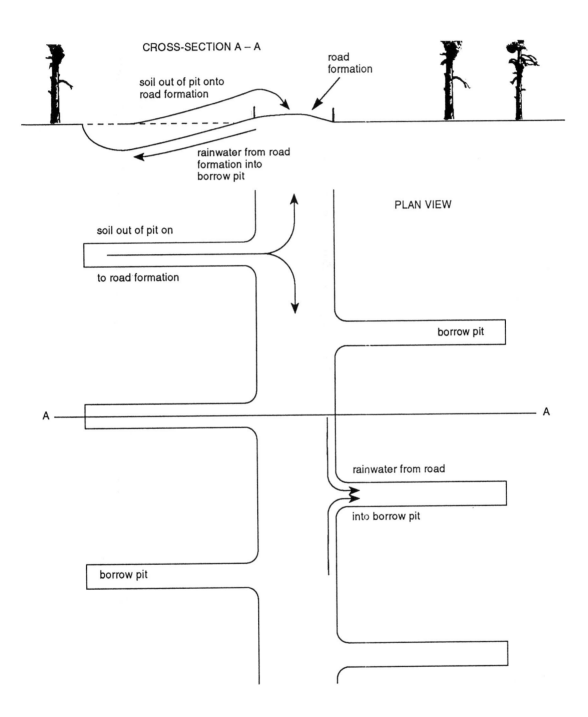

CROSS-SECTION A – A

soil out of pit onto
road formation

road
formation

rainwater from road
formation into
borrow pit

PLAN VIEW

soil out of pit on

to road formation

borrow pit

rainwater from road

into borrow pit

borrow pit

DIAGRAM 32   Use of borrow pits to facilitate soak-away drainage on an incline

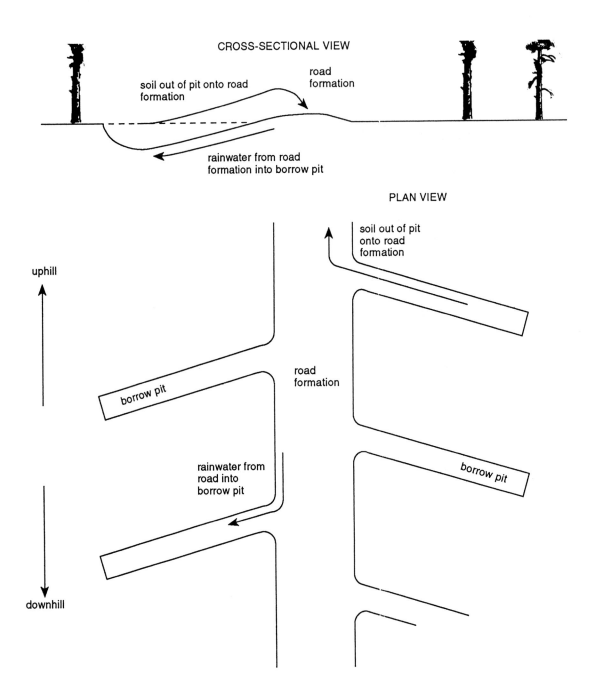

CROSS-SECTIONAL VIEW

soil out of pit onto road
formation

road
formation

rainwater from road
formation into borrow pit

PLAN VIEW

soil out of pit
onto road
formation

uphill

borrow pit

road
formation

rainwater from
road into
borrow pit

borrow pit

downhill

DIAGRAM 33   Use of borrow pits to facilitate soak-away on a sideslope

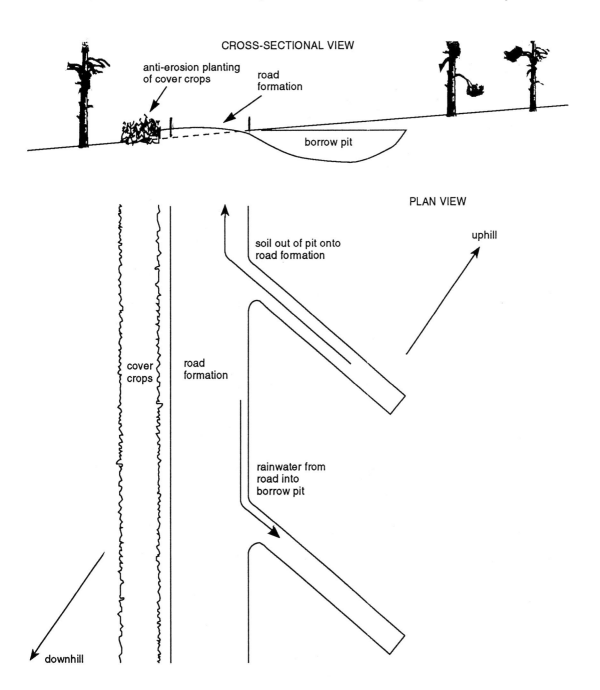

CROSS-SECTIONAL VIEW

anti-erosion planting
of cover crops

road
formation

borrow pit

PLAN VIEW

uphill

soil out of pit onto
road formation

cover
crops

road
formation

rainwater from
road into
borrow pit

downhill

the fines will act as a barrier to the percolation of the collected rainwater into the soil, and so effectively reduce the capacity of the pits to ameliorate the problem.

If sources of clay can be found in the region, clay can be mixed into the upper layers of the road formation to make a more satisfactory medium. Adding loose gravel without clay binder will not help matters much. Try dumping one tipper load of clay to every three or four loads of gravel, mix by windrowing with the grader, lay and compact. With observation of the results and experience, it will be possible to know whether this proportion of clay is adequate for the particular road being treated or if it should be increased.

When constructing a road straight up-and-down a slope, the pits should be placed alternately either side of the road, approximately at right angles or, better still, slightly inclined (see Diagram 32 on page 122), to facilitate any feeding of water along the road edge into the pits. This will also tend to result in any overflowing of the pits to occur at the end furthest from the road and so help to protect the road itself.

Where construction of the road is obliquely across an incline, the pits should all be on the upper side of the road, slanted away from the 90° alignment, as shown in Diagram 33 on page 123.

The pits should be slightly wider than the width of the blade of the bulldozer, or the bucket of the front-end loader, being used. They should be long enough for the operator to work comfortably if a bulldozer is being used. The depth of, and distance between, the pits will be determined by the amount of soil required to build up the formation, the distance over which water can be allowed to flow without scouring occurring, and primarily by the volume of water to be contained during a spell of rain and the rate at which it evaporates or percolates into the soil. The pits themselves may require protection against erosion starting around their steep-sided edges: it is advisable to plant suitable anti-erosion covers in a swathe a couple of metres wide around their rims.

Inspection before the commencement of the rainy season, and frequently for the period of its duration, should ensure that there are no obstructions to the free entry of water into the pits.

CULVERTS

Cross-sectional areas

Culverts that simply drain the contents of roadside drains from one side of the formation to another will only need small cross-sections in theory but in practice the smallest size used should be not less than $1/2$ metre in diameter (18 inches) for pipe culverts and $1/2$ metre × 1 metre for wooden culverts (but see *Other forms of culvert*, page 134). Smaller sizes are a nuisance to clear if they become blocked, which in the writer's experience, they do very frequently. However, when one is dealing not only with the water from the roadside but also with water from streams (albeit that they only flow during heavy rains), it is necessary to ensure that the total cross-section of any installation is more than adequate to allow for any partial blockage that may be caused by branches or other materials. It is worth while to walk up stream and river beds and look for evidence of past flood levels, driftwood lodged up on the banks, silting or other evidence of flows outside the normal channel. From such traces one can get some idea of what the cross-section of the river is when it is in flood. This can fairly easily be quantified in suitable places by measuring the depth of the water multiplied by the width of the channel. Check with local people on historical levels if possible and see whether they are in agreement with your own observations. When a figure has been obtained in which you have confidence, double it, or at the very least increase it by 50% to get the cross-sectional area of the culverting that has to be installed. This may sound over generous but remember that human activities in clearing forest and putting land under cultivation invariably increase runoff during heavy rains compared with natural forest conditions. You will need that degree of safety margin.

Culvert grids

The value of the cross-section can be easily reduced if the culvert becomes blocked. If land clearing will take place upstream and the stream carries driftwood down with it during floods, it is very worthwhile to protect the entrance of the culvert with a grid sloped from the stream bed upwards at 30 to 45 degrees from horizontal, NOT steeper, to a level above the culvert entrance. Driftwood will float up the grid as the water level rises and allow the free flow of water through the lower part of the grid and the culvert itself (see Diagram 34 on page 126). For wooden culverts, two round beams, one buried in the stream bed and one above the culvert, with stout grid members nailed to them at about a 30 centimetre (12 inches) spacing will do. For steel or concrete pipe culverts it is better to cast $5^1/_2$ centimetre ($2^1/_4$ inch) wide slots in the upper edge of the concrete facing and cast in a concrete foundation, also slotted to take the grid into the stream bed below its normal level. Then place sawn hardwood grid members, about 5 centimetres (2 inches) thick by 15 centimetres (6 inches) deep, into the slots to form the grid. This means that it is possible to lift the grid partly or wholly out of place when the accumulated driftwood is being removed or repairs to the culvert or to its facing are being effected.

Wooden culverts

Where durable timbers in log lengths greater than the width of the road formation are available, simple two-log culverts are easy to lay. They can be very crude, just two logs cut to length, dropped into a channel and covered with small logs put on top and covered over with earth (see Pictures 5.1 & 5.2 on page 136). More permanent structures can consist of capsils placed on dry stone walls or gabions, either decked directly with sawn or cleft lumber (see Diagram 35 on page 127), or overlaid with beams and decking and then earthed over – at which stage they are becoming more like bridges than culverts. Where the two logs form the capsils of stone or concrete foundations, do not forget to leave spaces under the cross beams for the use of spanners to tighten the holding bolts. The foundations will thus have a castellated appearance. If concrete is being used it is effective to bed the capsil into the top 3 centimetres ($1^1/_4$ inch) so that it fits snugly rather than to go to all the trouble of casting in mounting bolts.

The most essential thing in the design is to ensure that the dimensions of the outlet – both vertically and horizontally – are greater than the dimensions of the inlet, so that anything that

DIAGRAM 34   Driftwood grids for culverts

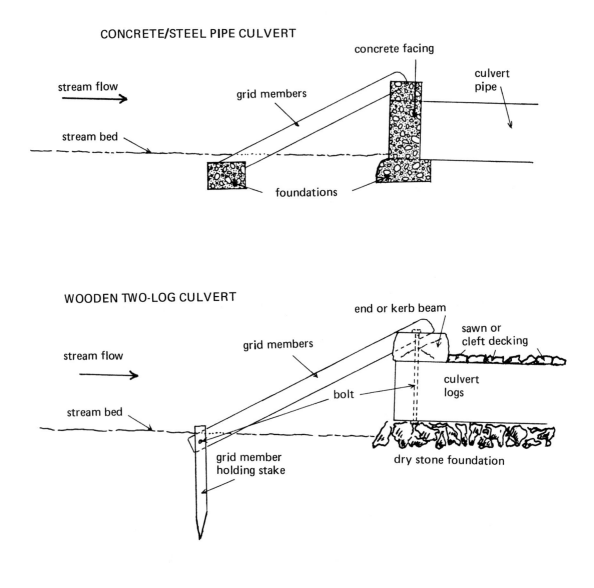

CONCRETE/STEEL PIPE CULVERT

concrete facing

culvert pipe

stream flow

grid members

stream bed

foundations

WOODEN TWO-LOG CULVERT

end or kerb beam

sawn or cleft decking

grid members

stream flow

bolt

culvert logs

stream bed

grid member holding stake

dry stone foundation

DIAGRAM 35   Wooden two-log culverts

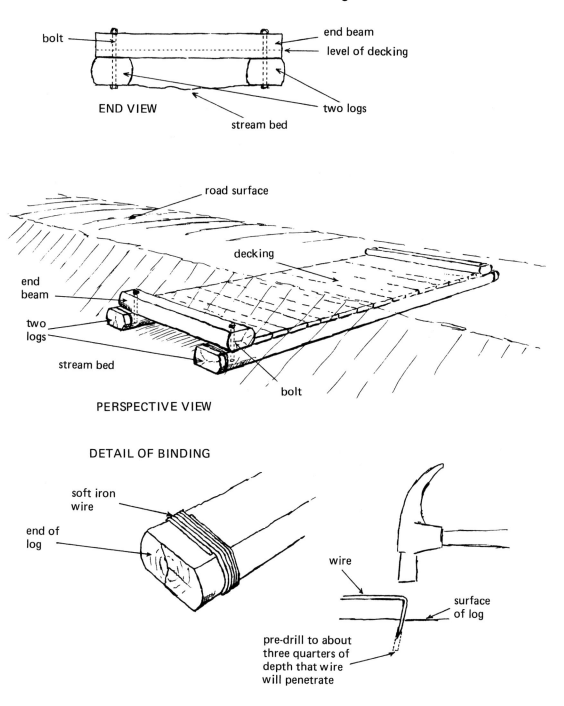

bolt

end beam

level of decking

two logs

stream bed

END VIEW

road surface

decking

end
beam

two
logs

stream bed

bolt

PERSPECTIVE VIEW

DETAIL OF BINDING

soft iron
wire

end of
log

wire

surface
of log

pre-drill to about
three quarters of
depth that wire
will penetrate

DIAGRAM 36   Earthing in a kerb beam for a two-log culvert

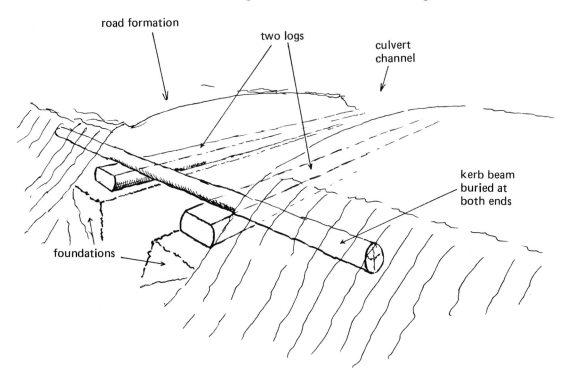

DIAGRAM 37   Profiling road to minimize flood damage to a bank of culverts in an alluvial area

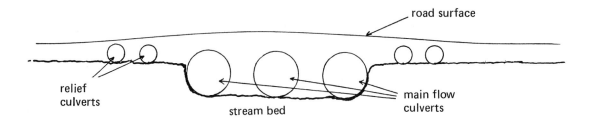

flows in can flow out. This is most important. Many smaller culverts of this type have become blocked for neglect of this very simple principle.

The great advantage of the two-log culvert and its derivatives lies in its integrity across the width of the road and therefore its ability to spread its load and resist destruction by uneven settlement of the formation. Neither concrete piping nor steel nestable culverting have this advantage. Indeed two log culverts are so reliable that they tend to get totally neglected, their very existence forgotten until, after very many years, subsidence occurs following erosion of the channel to the point at which the stream has completely undermined one or other of the two logs.

If a horizontal band saw is available the logs can be trimmed flat top and bottom, and if need be on the inside vertical face, before laying. This work can also be done with saw, axe and adze. It is wise to bind the ends with 8 or 10 gauge soft iron wire, about five or six turns well driven in (see detail in Diagram 35 on page 127) to prevent the ends from splitting open with time. The cross linking at the ends is best done either by bolting on two large diameter "kerb" cross beams, one at each end (which will also help to hold the earth covering in place), or by putting two longer kerb pieces across and burying their ends in soil so they cannot move (see Diagram 36 on page 128). If preparing bolts can take a long time, this is an effective method of getting the job done without delay and the bolts can always be added later.

The bolts should be thick enough not only to be strong but also to allow for corrosion. Large thick steel washers should be used under both the heads and the nuts. In practice bolts of 22 millimetres ($^7/_8$ inch) diameter placed into holes of 25 millimetres (1 inch) diameter will allow fitting without much trouble. Smaller tolerances are not easy to cope with in field conditions and may jam up.

The upstream kerbpiece can double as the upper grid beam (see *Culvert grids*, above). Between the kerb pieces, sawn or cleft decking can then be fitted in, often without being nailed or bolted in any way as the soil covering will prevent movement. The fitting of the decking does not have to be tight, indeed it is better that water should be able to drain through, and so small gaps are desirable. If the logs or capsils are more than about $1^1/_2$ metres apart it will be better, from the point of view of strength, to place and bolt beams across at a metre (40 inches) spacing between them and deck on top of the beams, the decking being placed in alignment with the two capsils. When laid this way the decking on the beams will have to be secured by nailing to prevent movement during earthing over. Provided that the deck laying extends over the two logs it will not be necessary to close the holes between the ends of the beams.

For these types of culvert, using the very hard durable tropical hardwoods (nearly always called "iron" woods), the logs will need to be 25 or more centimetres (10 inches) thick and 30 centimetres (12 inches) or more wide, the kerbs and cross beams 20 centimetres (8 inches) thick and the decking 7 to 10 centimetres (3 to 4 inches) thick for main roads. These sizes will ensure adequate strength and durability for many decades and, as the earth fill has of itself a considerable bridging effect which increases with its depth, there is really no point in limiting the depth of earth fill that may be placed over the culvert. The limit is, in practice, more likely to be set by the length of log available for the construction of the culvert since, as a rough rule of thumb, it must be as long as the full width of the road plus twice the depth of fill. For harvesting roads which tend to be more superficial, smaller beam thicknesses, about two thirds of those for main roads, will suffice.

Steel culverts

There is, on the face of it, not much designing to do with these as they come ready to bolt together. Decisions only have to be made with regard to the diameter of the culvert required and the thickness of its metal. These questions can be linked to that of whether to use several culverts of a smaller size or one of a large enough size to cope with the anticipated water flow. The shallower the fill, the greater the localisation of load experienced by the culvert when vehicles pass over it, so the thicker the metal of the culvert needs to be. On the other hand, small diameter culverts are inherently more robust and will survive with shallower fills than big ones.

In flat lands, covering over a single large culvert to provide the required cross-sectional area is not necessarily the sensible answer if a lot of soil would have to be found to ramp up the road either side of it. Also a cylindrical culvert does not have the wooden culvert's ability to remove shallow water quickly because of its cross-sectional shape. So where flat land is liable to flooding, or flash flows occur in shallow river beds, banks of smaller culverts in the stream bed may be the better answer to the problem (see Diagram 37 on page 128) with, perhaps, relief culverts either side and the careful profiling of the road away from the culverts to give further relief in extreme circumstances when the water level rises above the roofs of the culverts.

The culvert bed should be at least 25 centimetres (10 inches) deeper than the floor of the drain or stream and this extra depth filled with not coarser than 5 centimetre (2 inch) crushed stone or gravel to bed in the culverting. The channel containing the culvert should be at least 50 centimetres (20 inches) wider than the outer diameter of the culverting so that there is room either side to work in more crushed stone under the sides of the culvert when it has been laid. The culvert should be topped off at least to the level of the middle of the culverting (see Diagram 38a on page 131) with crushed stone, preferably to 25 centimetres above the culverting if this can be afforded (Diagram 38b). Finally, fill over with soil to a minimum of 1 metre (40 inches) for culverts up to $1^1/2$ metres (60 inches) in diameter, $1^1/2$ metres (60 inches) of soil for culverts of 2 metres (80 inches) diameter. If banks of culverts are being laid it is important to allow space between the individual culverts, preferably about $^1/2$ metre (20 inches), so that the gravel can be properly worked in between them (Diagram 38c).

Longitudinally the channel should be cambered a little so that the outfall is slightly steeper than the entry (Diagram 39a on page 132). This will counter any tendency for the soil of the formation to settle under the weight of traffic and should prevent the culvert from sagging (Diagram 39b). Sagging can enable water to lay in the culvert and cause the softening of the foundation over a period of time which, in turn, could lead to the eventual collapse of the culvert.

The bottom plates of the culvert should be laid from outlet towards inlet so that the water can flow over them like water flows over the tiles on a roof. Some steel culverting has right-angled flanges to enable the upper and lower halves to be bolted together. Other types are slotted to enable the plates to interleave with one another when they are assembled and are then drawn

Bedding in concrete/steel pipe culverts

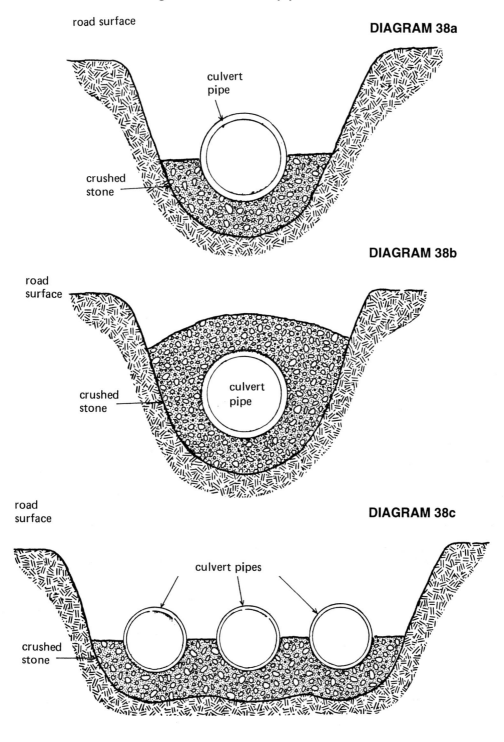

Bedding in concrete/steel pipe culverts

DIAGRAM 39a

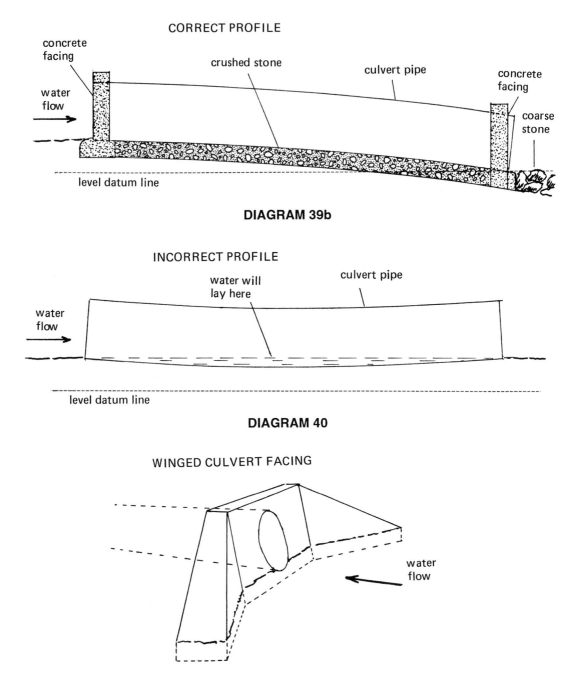

CORRECT PROFILE

concrete facing

crushed stone

culvert pipe

concrete facing

water flow

coarse stone

level datum line

DIAGRAM 39b

INCORRECT PROFILE

water will lay here

culvert pipe

water flow

level datum line

DIAGRAM 40

WINGED CULVERT FACING

water flow

tight either with staples and a bar to bend them in place or bolts and eyes to draw them tight. End pieces cut at 45 degrees or half plates to enable the pipe to terminate with a neat 90 degree end can be obtained but it is perfectly satisfactory – if not so pretty – to allow one more plate on the bottom than on the top (since the plates are designed to bridge the opposing gaps during assembly) and end the pipe with the channel slightly longer than the arch.

When earthing over culverting of more than 1 metre (40 inches) in diameter, it is desirable to prop the culverting up to prevent it from becoming flattened. The simplest way to do this is to lay a plank in the bottom of the culvert that is about 15 centimetres (6 inches) wide by about 5 centimetres (2 inches) thick; cut some short planks about 30 centimetres (1 foot) long by 10 centimetres (4 inches) wide and $2^1/2$ centimetres (1 inch) thick and an equal number of props that will be (with the thickness of the planks added) about 1 centimetre greater for every metre diameter of the culvert than the diameter of the culvert being laid. The props are then stood on the bottom plank and, with the short top planks on the top of them, hammered into an upright position at about $1^1/2$ metres (5 feet) apart down the length of the pipe. This will force the pipe into a slightly oval shape with the greatest axis vertical and prevent the weight of the soil during filling from crushing it. Remove the props when the fill has been consolidated.

Where the culverts are likely to be subjected to strong currents of water, the upstream side at least should be faced with concrete or masonry, much as is shown in Diagram 40, even if a grid is not necessary. Ideally the downstream side should be similarly treated if there is the money for it. Alternatively, well laid dry stone walling using large shards of stone "as blasted" or well shaped sedimentary rocks, or the construction of gabions over and around the face, are effective. Where concrete is used the culvert's end channel plate can be bedded into the foundation when it is poured. This will ensure a good seal on the underside where it is most needed. A winged facing may be necessary (Diagram 40) but bear in mind that unless reinforcing metal is used, the bigger the structure the more it will cost and the more likely it is to crack with time. It is better to do only as much cement work as is required to protect and seal the entrance of the culvert and, having shaped the soil filling to reduce the chance of it slumping, to face it with stone or netting and/or plant covers to prevent erosion. On the downstream side the risk is that water falling out of the last of the culvert channel plates will, in time, dig a hole which will eventually undermine the culvert at that point. The cheapest method of dealing with this is to allow the water to dig up to half a metre in this way and then to place large heavy stones into the hole, working the first few well under the end plate, so that the water from then on dissipates its strength on the stones and further undermining is prevented.

Concrete pipe culverting

Much of what has been written above on the subject of steel culverting applies also to concrete pipe culverting. However, concrete pipes are less "good tempered" than steel culverting. As with steel plate culverting, the pipe joints must be laid the right way round, male joints facing downstream, female joints upstream, and great care must be taken to prevent damage to the joint rims. When using the larger pipes or the longer culverts required for main roads, and if the bed does not consist of firm rock or hard clay, it is desirable to compact the channel bed and lay the pipes on a foundation of concrete containing reinforcement rather than just use crushed stone. This is to reduce the risk of sagging and cracking. If a reinforced concrete foundation is used, then the pipe joints can be precemented on the outside before the pipes are covered in. This is to reduce the possibility of leakage at the joints resulting in cavitation of the fill. Fill over firstly with a layer of crushed stone, and then with earth. If a concrete foundation is not used it is probably better not to precement the joints but to pay great attention to the laying of the culvert in crushed stone, ensuring that it is well done with an adequate camber and, if the pipes are big enough to allow access, to cement the joints during a dry spell from the inside, a year or two after laying when any settlement has occurred.

Other forms of culvert

If the project has a sawmill and plenty of durable interlocked* grained timber for sawing, simple "four-plank culverts" can be made with planks cut to the length of the culvert required, about 4 centimetres ($1^1/4$ inches) thick. For each culvert the top and bottom planks should be 30 centimetres (12 inches) wide and the side planks 25 centimetres (10 inches) wide, two of each per culvert. The narrower planks are stood edge up on one of the broad planks at either edge and the other broad plank is placed on the top to form a box sectioned pipe. Nails are driven through the edges of the broad planks into the perpendicular narrower planks which form the sides of the culvert and the whole unit is then bound with 8 or 10 gauge soft iron wire, one binding at each end and one binding about every $1^1/2$ metres between. The culvert hole will measure 25 centimetres (10 inches) vertically by 22 centimetres (9 inches) horizontally. Being smooth bored it will be easy to unblock. Such easily mass produced culverts are useful for the smaller roads that have been side cut into hillsides where numerous cross drainage culverts, closely spaced, are necessary to reduce roadside erosion.

Culverts are now being manufactured of plastic in several countries, including Canada and the United Kingdom. They are available in sizes up to 1.5 metres (60 inches) in diameter (see Picture 5.6 on page 138). They are light, laid in one piece and extremely durable, being unaffected by acid conditions that would destroy a steel culvert. They do have to be laid and bedded-in with care to avoid damage, but once properly in place are permanent.

A number of tropical countries are developing industries as well as agriculture and forestry. It is as well to see what plastics are produced locally and whether they may be relevant to the problems of culverting in the smaller sizes at least. Pipes of waste polythene in the larger diameters, 25 centimetres (10 inches) and above, or spirally produced piping, which may be available up to around 45 centimetres (18 inches) in diameter, are so smooth bored that they are unlikely to get blocked except at the entrance. They usually come in standard lengths of around 7 metres (23 feet) which make them very suitable as culverts to relieve the roadside drains on side cut roads or to help drain junctions. They can be quite cheap, light to carry and easily cut

---

* Tree growth can result in grain parallel to the centre of the log. Such timber is easily cleft along the grain. Many tropical trees produce grains that spiral around the centre of the log, changing the direction of the spiral annually. This results in an uncleavable cross-ply structure. This is called interlocked grain.

to length. Sleeves may be available to enable them to be joined together for longer lengths. Spiral galvanised steel piping in the smaller culvert sizes may also be available to order, and be suitable for the same sort of purposes, but beware, the bores of some types of steel spiral piping are not as smooth as plastic pipes and blocking can be a problem.

Culverts can be made of all sorts of other things, from oil drums to reinforced concrete, from railway sleepers to masonry arches. Basic principles remain the same: what can float in must be able to get out; foundations must be firm enough to bear the overfill and traffic of the road; the water must not be allowed to erode holes that can result in subsidence and the durability of the structure is very important because digging out and replacing failed culverts can be very expensive, especially if they are buried really deep down.

Picture Group 5:   Simple wooden bridges, and wooden and plastic culverts

5.1/5.2   Loggers' culverts: made very quickly and with remarkable dexterity out of whatever timber is at hand, they are effective in the short term but, if buried deeply, can be a considerable embarrassment later, when they rot, to anyone who wishes to repair and continue to use the road. In the top picture skilled bulldozer operators feed logs across and lay them to make a crude decking that in the lower picture is being covered with earth.

5.3/5.4 Loggers' bridges: crude, quickly made and effective for short term usage. They too, like loggers' culverts, soon rot and collapse because they are made of whatever wood lays readily to hand. In the top picture the abutments – made with long logs laid back into the banks and earthed over to make them stable, then placed in layers across logs parallel to the river's edge – support the first two or three beams which have been put in position to span the river. In the lower picture all the beams have been put in place and the bridge is ready to be earthed over. The replacement of such a structure with a permanent bridge later requires that the whole of the abutments be removed first.

5.5   A hollow log culvert being put into place in a low lying piece of ground planted to rubber. The harvesting road was later built up on an embankment that covered the log by some half a metre at the edges, and rather more in the centre. Installed in 1956, the log still served its purpose in 1986. It was burnt out as described in the text, trimmed to the correct length and towed to site by the small bulldozer in the picture. Since this picture was taken, the rubber has been cleared and the rather waterlogged land planted, more suitably, to oil palms.

5.6   Large-sized plastic culverts up to 1.5 m diameter are now becoming available. They are durable, light and very easy to install, but they come in long lengths and this can provide transport problems when comparing them to steel culverts, which are made of many small plates. (Photograph courtesy of Mr John Anstey of the Forestry Commission, UK.)

5.7   An estate main road culvert under construction. Hard durable timber on dry stone walling made from the fragments of rocks blown-up to clear the stream bed. Note the cleft decking laid out for nailing in place prior to earthing over the whole structure.

5.8   These pictures were taken in 1957 and the culvert is still in service. No repairs have been effected since it was built.

5.9/5.10   The culvert in the process of being earthed over. The felled land was planted to rubber for twenty-four years and is now under oil palms. The land opposite was planted to oil palms for eighteen years and is now under cocoa. This underlines the need for estate roads and road related structures to be designed and built to last.

5.11/5.12 Gabions make good culvert foundations and facings for bridge abutments as well as walling to control unstable land. These are made of concrete reinforcing mesh and fifteen centimetre stone ex-quarry. They have been beautifully made and correctly laid, like bricks, and tied in together to form a unitary but flexible structure.

5.13   The simple hand winch pile driver shown in Diagram 45 on page 152 is pictured here being used to construct a wharf for a cocoa estate. In the upper picture the hammer has been raised for the first blow on a pile that has just been positioned for driving.

5.14   In this picture the release rope has just been pulled and the hammer has fallen. Note the temporary scaffolding made of poles from the jungle which supports the piledriver and the weight of piles being lifted into place. Note also the stabilising guys running from the head of the rig to the far side of the scaffolding.

5.15 A line of five piles driven for the construction of a bridge on a main estate road in 1957. The sixth pile is still against the piledriver. Railhead rings can still be seen in position on the fifth and sixth piles.

5.16 A bridge protector alongside a bridge over a river known to be liable to flooding and carrying heavy loads of driftwood – see next picture …

5.17 In this picture, taken several years later, the morning after heavy rains and widespread flooding the bridge is seen to be intact. The water level reached to the dark mark on the road surface halfway between the bulldozer and the car.

5.18 The protector, almost invisible under its load of driftwood, took the strain and allowed water to flow freely underneath. As the flood level fell the driftwood settled back onto the protector. It was cleared with winch and chainsaw during the day.

5.19 This bridge on a government road, itself a replacement for an earlier bridge destroyed by floods (see the pier standing unused to the left of the bridge), was destroyed in the same floods that the bridge in the previous pictures survived.

5.20 Wooden bridge pier construction where piling is not possible. Note concrete foundation, with wooden cap, supporting fine hand-hewn uprights with cross-bracing, capsil on top carrying beams under the deck in sets of three immediately under the deck runners. The main span is 10 metres, total bridge length 22 metres, maximum load 10 tonnes.

BRIDGES

General principles

The selection of bridge sites has been covered briefly earlier on in the book (see page 19). Sites can be classified by whether they are on the river's torrent course or flood plain course, whether the banks are of rock or of soil, whether one span or several spans are required and, if several spans, whether the river bed is soft, stony or of rock.

Another factor which may affect the approach to building a bridge is whether it is necessary to have the road always open or whether some degree of interruption can be accepted. If the latter, either a low-level bridge or a ford may be acceptable.

The materials, building skills and handling equipment available will have an overriding influence on the designs of bridges. Good strong durable tropical hardwoods are, without doubt, the finest and most practical principal raw materials for estate bridge building. They can be supplemented by other materials if necessary, an example being the use of stone and cement for foundations if the ground is too hard for piling. In this book, therefore, it is the intention to concentrate on simple wooden bridges in the belief that the construction of steel and reinforced concrete spans for a bridge, as distinct from the casting of abutments and piers, will be more than project staff can cope with without outside assistance. Some consideration will be given both to recently developed designs for wooden truss bridges and to stone arch bridges, many of which were built by medieval constructors with neither sophisticated equipment nor advanced engineering knowledge, and which survive to the present day. In some parts of the tropical world, the prodigal destruction of the natural forest means that the strong, durable hardwoods are no longer available and this creates a problem particularly where piling is necessary. Some ideas as to how this problem may be overcome are discussed later (see pages 172 to 175).

Site type will largely determine the foundations to be used. If the site is rocky, piling is impossible. If the bridge is to cross a deep narrow valley it is likely that the bridge foundations can be placed well above any conceivable water level and piling will be unnecessary. For such bridges the preparation of foundations can be as simple as those for wooden culverts (see Diagram 35 on page 127) described earlier, a bridge on such a site is in a sense merely an extension of a wooden culvert structure. If the foundations are liable to be washed over by the river when it is in flood, something more substantial like gabions or a reinforced concrete footing will be required. (See Picture 5.20 on page 145.) It is preferable that this foundation should be built up to capsil level – at the very least it should extend up above flood water level before the wooden structure starts, to ensure that the bridge can withstand the assaults of driftwood. Where interspan piers are involved piling is the very best answer if piles can be driven. They have immense lateral resistance to withstand the flow of the water and can be driven with either a piledriver on a scaffolding or on a pontoon. A bridge protector can be provided (see Diagrams 41 and 50 on pages 149 & 162) where the river is known to bring down large quantities of driftwood. Where massive rock forms the bed of the river it will usually be possible to prepare a footing (with explosives if necessary) on which reinforced concrete can be cast during the dry season to make a pier to capsil level. If driftwood is a problem the pier should be shaped on the upstream side as shown in Diagrams 41 and 49. The most difficult sites are those in which the river beds contain rocks deposited by the river that are big enough to prevent piling but, not being bedrock, are not suitable to cast a concrete foundation on. If there is a reliable dry season during which flow is negligible it may be possible to dig out a trench in the bed and cast a reinforced concrete foundation upon which a reinforced concrete pier can be built. Such a foundation needs to be massive and deep in order to withstand floods. It will be expensive. It may be cheaper and better to look for an alternative bridge site.

Avoid building bridge abutments that in any way constrict the river. To do so is to invite the river to deepen its bed between the abutments and probably undermine them with possibly catastrophic results. If the bed is of hard rock the constriction will result in the unnecessary raising of the water level upstream of the bridge and very strong currents under it during floods which will endanger the bridge in the long term. It is better to spend money on additional spans than on revetments, wings and all the usually advocated means of protecting abutments that are too close together for comfort.

Bridge widths should ideally conform to road widths since road widths are going to be related to traffic density. A single lane width bridge is logical on harvesting roads, provided that it is wide enough for all the vehicles that are going to use and maintain the roads to pass over it with reasonable clearance. For main roads, width enough for vehicles to pass one another is very desirable and, if suitable timber is available on the estate just for the cutting, the additional cost of making a bridge wide enough for two lanes is far less than twice the cost of building a single lane bridge. It is both cheaper and structurally more satisfactory to make the thing big enough in the first instance than to have to enlarge it later (see *The long term view* page 1). If reinforced concrete or steel bridges are to be built by a contractor, then the expenditure on a two lane width bridge may be too high to be economically viable and single lane widths and traffic restrictions may have to be accepted. Both those media are more amenable to subsequent bridge widening than timber.

It is desirable that multi-span bridges, particularly if on piles, should not be on corners but should be straight themselves and have several tens of metres or so of straight road leading onto and off them. A bridge on a curve is much more complicated to build and operate as side stresses are incurred when vehicles travel over it at speed and this can involve both extra wear and the need for super-elevation of the decking. With a simple wooden bridge, skewed construction of single span and two span bridges is feasible up to about 15 degrees from square, but even slight skewing introduces considerable complication when applied to multi-span, truss or other more sophisticated bridge forms. It is better, as far as possible in such cases, to build one's bridges square with the alignment of the road and to re-align the road to facilitate this if necessary.

Simple wooden bridges (see Diagram 41 on page 149 for names of bridge parts)

Piles

Driving wooden piles is not something to be frightened of. It may be possible to hire a piledriver or to get a contractor to drive piles for bridges and jetties. Failing that it should be within the capability of the average estate carpenter to make an effective piledriver capable of handling piles of up to 35 centimetres (14 inches) diameter and 8 metres (25 feet) long, which will cover the needs of most projects (see Diagram 45).

Logs within a species may have either straight grain or interlocked grain. Even within one tree it is possible to have some lengths that are of straight and others that are of interlocked grain. It is impossible to cleave interlocked grained wood cleanly but it is easy to cleave any wood that has straight grain. For decking and beams, and to some extent for capsils, straight grained wood is acceptable, but for piles, the quality that makes straight grained wood easy to cleave also makes it split readily under the piling hammer. Logs for piles have therefore to be very carefully selected to be of interlocked grained wood, to be very straight so that they will not "spring" under the hammer, and to be devoid of large knots, hollow hearts or other defects. Only woods classified as very durable should be selected for piling (see Tables 4 & 5 on pages 169 to 171). If the water in which they are to be driven is salty, selection for resistance to teredo will also be necessary. If possible, piles should be of the right diameter without being sawn except, perhaps, to trim off an excessive thickness of sapwood. Sawn baulks will bow after sawing (see Diagram 42 on page 150) which means that they will spring under the hammer unless they are very short. They are therefore not normally suitable for piling.

Pile size will not be dependent solely upon strength, the larger the diameter the longer it will last and the greater the bearing it will get from the ground it is driven into. Wood is generally very strong under compressive loads and so a pile that could be theoretically adequate to take the design load of a bridge can also be far too small to be practical. It is therefore as well to put in good thick piles that can be reasonably expected to stand up to all that the river can throw at them, driftwood, rocks, boats and the like. Make them at least the same sort of diameter as the capsils and beams that will be placed upon them.

Piles have to be prepared carefully. The point should be tapered four-square to a tip about $2^1/2$ centimetres square (see Diagram 43 on page 150). This tip may be bevelled, but should not be pointed because a fine point is liable to break if a stone is encountered, and this may result in the pile running crookedly. If the ground into which it is to be driven is known to be stony, a mild steel sheet sheath can be cut and welded together in the workshops and fitted over the point to prevent any risk of the pile splitting. Round points tend to wander badly. Three-sided points are difficult to cut to perfection in the field, they are really only of advantage when driving small piles where a mat of roots will be encountered in the first metre or so of soil – under such circumstances the very sharp edges can help initial penetration. The tops of the piles have to be crosscut absolutely square, then trimmed to a circle and a steel band fitted over to prevent the timber splitting under the hammer (refer to Diagram 44 on page 151). The best material for such a ring is old light-gauge railway line, the base of which has been discarded and the head of which is rolled or beaten by a blacksmith into a ring and welded. If railway line is not available, steel bar about $2^1/2 \times 3^1/2$ centimetres ($1 \times 1^1/2$ inches) can be used but will need to be rounded off on the inner corners to make fitting easier. The size of the ring should not be greater than the size of the hammer face otherwise the ring may ride up the pile and be shed when driving is in progress. Any knots or projections from the pile should be trimmed off before it is driven as they may damage the piledriver frame.

DIAGRAM 41 The names of bridge parts used in this book

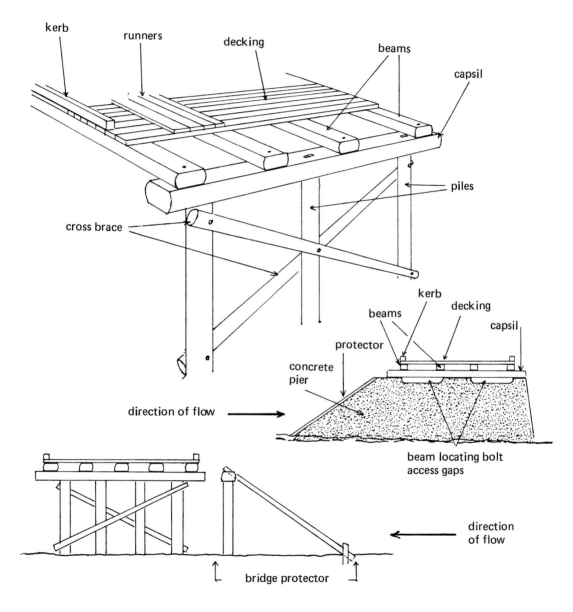

DIAGRAM 42   Milling timber for bridge construction

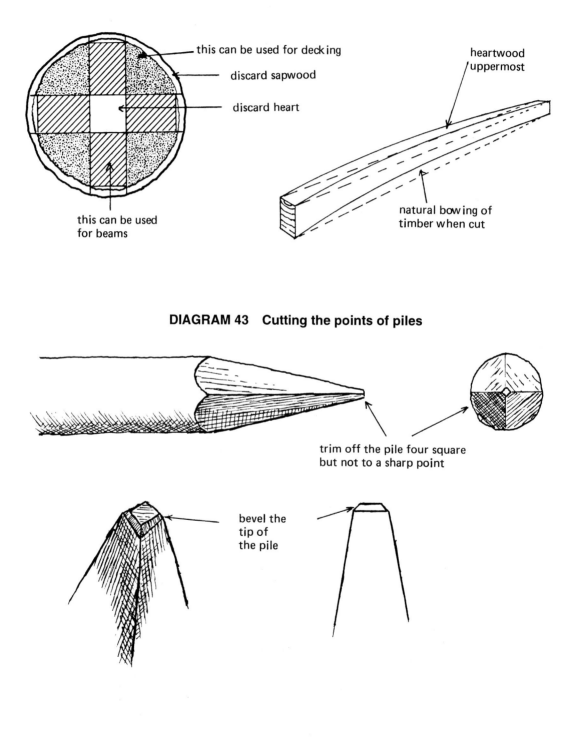

this can be used for decking

discard sapwood

discard heart

heartwood uppermost

this can be used for beams

natural bowing of timber when cut

DIAGRAM 43   Cutting the points of piles

trim off the pile four square but not to a sharp point

bevel the tip of the pile

DIAGRAM 44   Preparing the tops of piles for driving

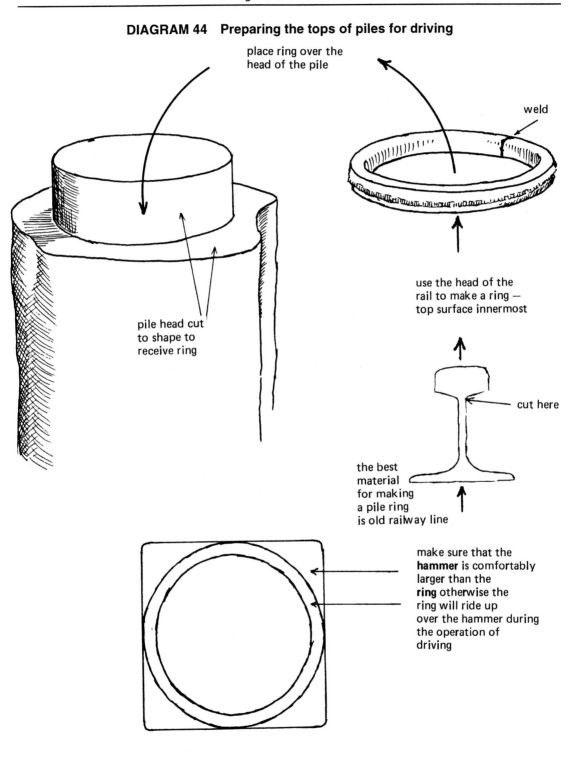

place ring over the
head of the pile

weld

pile head cut
to shape to
receive ring

use the head of the
rail to make a ring —
top surface innermost

cut here

the best
material
for making
a pile ring
is old railway line

make sure that the
hammer is comfortably
larger than the
ring otherwise the
ring will ride up
over the hammer during
the operation of
driving

DIAGRAM 45   A simple wooden framed piledriver

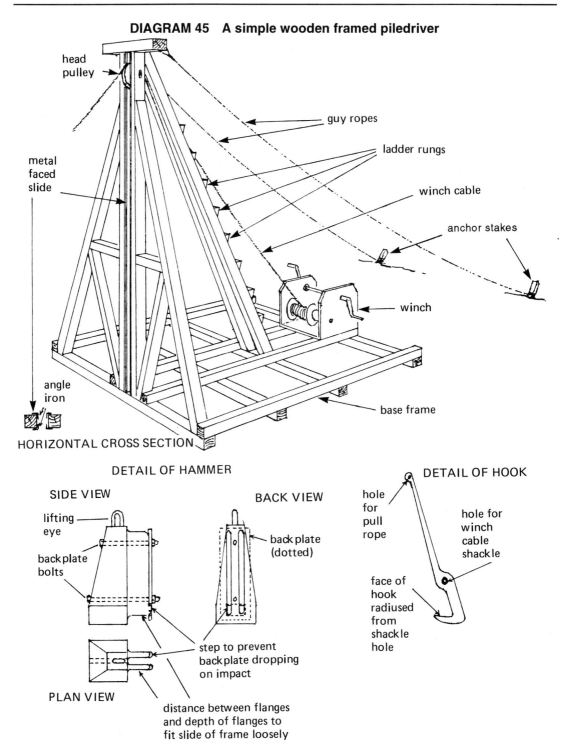

head pulley

metal faced slide

angle iron

HORIZONTAL CROSS SECTION

guy ropes

ladder rungs

winch cable

anchor stakes

winch

base frame

DETAIL OF HAMMER

SIDE VIEW

lifting eye

backplate bolts

PLAN VIEW

BACK VIEW

back plate (dotted)

step to prevent backplate dropping on impact

distance between flanges and depth of flanges to fit slide of frame loosely

DETAIL OF HOOK

hole for pull rope

hole for winch cable shackle

face of hook radiused from shackle hole

A simple piledriver is shown at work in Pictures 5.13 & 5.14 on page 142. This unit can be made up on an estate out of sawn lumber; a hand winch complete with ratchet, capable of lifting up to five tons, of the type used on many work boats; a pulley; some angle iron, cable, clamps, a shackle, nuts and bolts and one cast iron 500 to 750 kilogram ($\frac{1}{2}$ to $\frac{3}{4}$ tonne) hammer. Sketches of the whole unit and particular parts are shown in Diagram 45 on page 152. Assemble the unit with bolts, nuts, washers, plates and angle plates so that it can be dismantled easily for transport between jobs or storage between operations. Lay planks as a flooring (not shown in the diagram) on the base frame either side of the winch for the winch operators to stand on but do not fasten them permanently. The size of the piledriver can be varied to suit the needs of the project. In use it is mounted on a scaffolding of poles from the forest nailed together with 4 inch nails. The height of the scaffolding should be about the same as the height at which the capsils will be situated, and the height of the piledriver pulley must be great enough to allow the pile to be stood upright where it is to be driven while permitting the hammer to be raised and placed upon the top of it. The difference between the two heights will limit the depth to which the pile can be driven. Something in the order of five to seven metres should be satisfactory for most estate works. The piledriver can be used to lift capsils into place and to assist with the placing of beams on top of the capsils.

The piledriver is pushed into place with poles and crowbars and, when precisely situated, the head guys are secured back at about 45 degrees from one another to guard against the whole unit being up-ended if a very heavy pile has to be lifted. Once lifted into place, the pile is tied by ropes to the piledriver's slide before the cable is disconnected. The hammer is then hauled up and deposited on top of it and its backing plate secured before it is unshackled and the hook put in place. The operation is then ready to begin. The hammer is lifted, the rachet pawl put in place to prevent the winch from moving and the hook cord is pulled allowing the hammer to free-fall onto the pile. With the pawl released the hook can be pulled down and placed into the lifting eye on the hammer ready for the next lift.

This process is continued and after each blow the pile is scribed or pencil-marked level with the foot of the piledriver. This will show, blow by blow, the rate at which the pile is being driven as well as the total penetration. The height of the drop of the hammer controls the strength of the blow. While the pile is being driven easily, quite high drops can be used but, as it firms up, it will be necessary to reduce the drop to avoid shattering the pile. For a pile about 30 centimetres (12 inches) in diameter, the drop should be limited to about a metre and a half (60 inches) with a 500 kilogram hammer (less with a heavier one). Once the pile moves less than 3 millimetres ($\frac{1}{8}$ inch) per blow at that height, driving should cease. During driving, the ropes holding the pile in place will frequently need adjusting and it may be necessary to take some action to keep the pile straight if obstacles are encountered.

It has to be accepted that it will not be possible to drive every pile perfectly. Some will wander and some may refuse because they have hit a rock or long buried tree trunk. One's plans for a perfectly spaced row of piles may not always therefore be realised. One then has to improvise and drive extra piles in the next best place to achieve what is required. This can result in the need for a longer capsil than planned, so do not cut the capsil until the piling is completed. It is advisable to drive the piles in groups of four or more per capsil. This is not just

DIAGRAM 46a

DIAGRAM 46b

DIAGRAM 46c

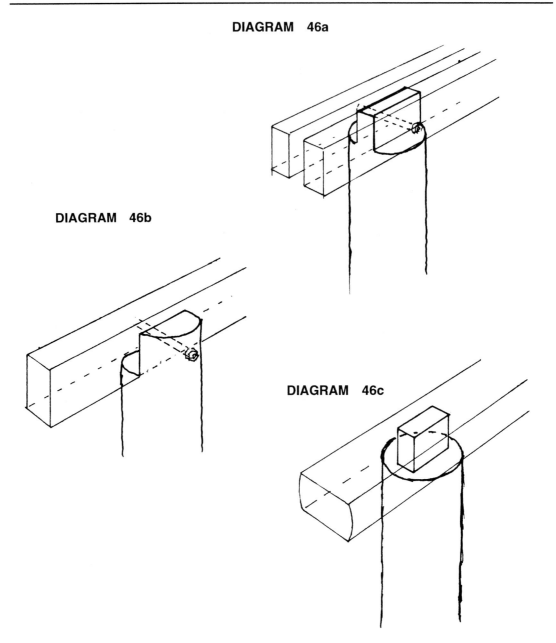

Various ways of fitting capsils onto piles

Sawn lumber capsils are shown in Diagram 46a and 46b. A log trimmed top and bottom either by sawmill or by axe, adze and saw, is shown as the capsil in 46c. Provided the capsil shown in 46c is lashed at the end to prevent splitting, the tongue can be wedged and no holding bolt is necessary.

DIAGRAM 47a   Aligning pile tongues

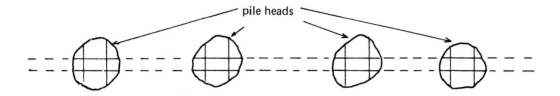

DIAGRAM 47b   Marking out positions of saw cuts to make tongues on the tops of piles

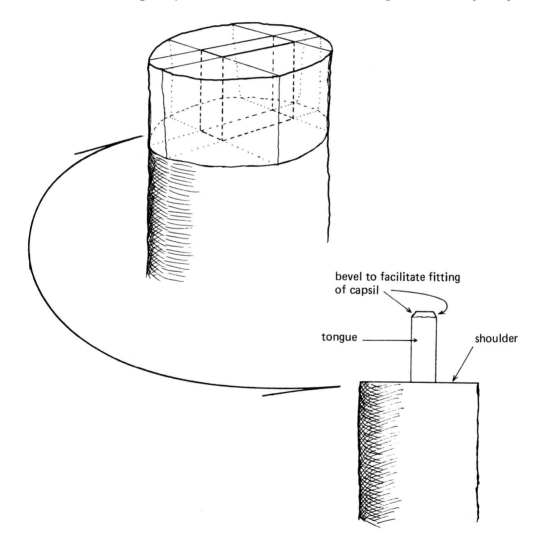

a matter of strength but also so that if one pile fails for any reason in the future it can easily be replaced without the bridge becoming unstable. The piles themselves will probably vary a little in size and strength: put the strongest and thickest of the group on the upstream side as it is always the upstream pile that gets clobbered by driftwood, hit by boats or suffers any other damage most frequently. At the same time as the bridge piles are driven it is as well to drive the bridge protector piles if a protector is required (see *Bridge protectors* page 169).

Next the piles have to be prepared to receive their capsils (see Diagram 46 on page 154 for different types of capsil mounting). Firstly stretch a line across them at the height of the top of the capsil. The piles are then all cut off level at this point. Next the line is laid across the tops of the piles to mark the sides of the tongues (see Diagram 47a on page 155). This should get all the tongues in a straight line and will indicate their thickness. The width of the tongues is then marked off. The piles are next marked at the height of the bottom of the capsil and the mark is made round the whole circumference of the pile. Vertical saw cuts are made to the thickness and width of each tongue and to the depth of the thickness of the capsil and horizontal cuts in from the circumferential mark are made to release the waste wood and leave the tongues standing proud (see Diagram 47b). The top edges of the tongues should be bevelled to enable the capsil to be fitted easily.

When wooden bridges fail due to rotting of the timber, the failure point is almost always where the joints are made. It has been the writer's practice to cover the tongue and the top of the pile with hot tar and to lay a couple of layers of tarred sacking on the shoulders of the piles before placing the capsil over them. Bridges built this way three decades ago still show no signs of rot at the joints.

Beams and capsils (see Tables 4 & 5 on pages 169 to 171)

In a bridge for an estate one looks for adequate strength and for durability and attempts to attain these desirable characteristics at the lowest cost. Many good bridges have been built without recourse to the extensive calculations of the type that professional engineers become involved in but bear in mind that the engineer's calculations are very much to do with the saving of expensive materials such as steel, concrete or imported sawn lumber whilst maintaining adequate standards of utility and safety. On a project where logs for bridge building are available from forest, which has in any case to be cleared for the planting of the project's crops, the need to be sparing in the use of these raw materials is less urgent and one can, with advantage, be liberal with their use if this will ensure reliability.

The equation below is important as a check to see that beams and capsils will bear the loads to be placed over them, *its objective is to ensure adequacy, not to save on timber*:

$$\frac{w \times l}{4 \times s} \leq \frac{b \times d \times d}{60}$$

where: w = maximum static load in kilograms at the centre of the beam between supports
         b = breadth of beam in centimetres

d = depth of beam in centimetres
l = span between supports in metres
s = timber strength class factor
the figures "4" and "60" are constants.

The left hand side of the equation establishes a figure based on the load to be passed over the beam and the distance to be spanned by it, whilst the right hand side of the equation enables one to try various cross-sectional dimensions to gain a result that must be equal to or greater than the value resulting from the left hand side. The use of the values shown enables adequate comparisons to be made in figures that are easily manageable (see example below).

CAUTION:

(i)   The timber strength class factors ("s" in the equation above) are based on tests carried out on test pieces selected to be heartwood of straight grain and good quality. Therefore when choosing logs for use as bridge beams or capsils only, the most perfect logs are acceptable in order to be comparable. Hollow, fractured, knotty or otherwise faulty logs must be discarded ruthlessly and sapwood discounted.

(ii)   The figure resulting from the use of this equation does not include a safety factor other than that the strength class factors given are conservatively rated.

As an example of the use of the equation let us ask the question: "What size beam of strength class 1 timber (see Table 4 on page 169) is required to carry a ten tonne load over a span of five metres?" Applying the equation:

left hand side

$$\frac{10,000kg \times 5m}{4 \times 25} \qquad = \quad 500$$

right hand side

$$\text{try} \quad \frac{30cm \times 30cm \times 30cm}{60} \quad = \quad 450$$

which is too low,

$$\text{try} \quad \frac{30cm \times 35cm \times 35cm}{60} \quad = \quad 612$$

which is unnecessarily high,

$$\text{try } \frac{25\text{cm} \times 35\text{cm} \times 35\text{cm}}{60} = 510$$

which is adequately close to and still in excess of the minimum acceptable figure of 500.

The answer is therefore that a beam which, as prepared to go onto the bridge capsils, contains a core that is not anywhere less than 25 centimetres wide and 35 centimetres deep along its entire length, will carry the load of ten tonnes. Referring then to Diagram 48 on page 160 (where logs from the forest that have only been trimmed top and bottom are shown), it is obvious that if those core dimensions are contained within the cross section of the top of the beam (excluding sapwood) – since everywhere else the beam will become bigger as one progresses to the butt – the beam will carry the load. There is no practical point in cutting off the excess material – nothing is to be gained by so doing and the operation will involve unnecessary work and expense. If fully sawn beams are being used (quarter sawn as shown in Diagram 42 on page 150) they should be sawn to be not less than 25 centimetres by 35 centimetres. To provide a safety factor of 100%, because the intent when constructing a bridge is to cater for a moving load, not the theoretical static load of the equation above, provide two beams of that size (the preferred method for sawn beams) or double the beam breadth (the more practical method with trimmed log beams). Therefore, to extend this example, four sawn beams 25 centimetres by 35 centimetres arranged in pairs or two 50 centimetre by 35 centimetre trimmed log beams, one under each of the running tracks, will carry a twenty tonne gross weight lorry across in safety. This then is the skeleton of the bridge. Additional beams will be required to support the decking in the centre and at the edges to make a practical structure.

The generous use of timber when calculating the sizes of beams, capsils, piles and other bridge parts will not only contribute to safety but also to durability as a great deal of timber then has to rot before a bridge will become too weak to do its job. It will be found in practice that by combining generous scantlings with the most durable of tropical timbers it is feasible to earth over a short bridge and gravel it so that a grader maintaining the road does not have to stop or change its mode of working when it comes to or leaves the bridge. For long bridges this is less practical because the earth will suffer "dry-breakdown" in a long dry spell and disintegrate under traffic.

It has been noted that when using logs rather than sawn beams it saves work if the log is flattened on the top and bottom and not on all four sides, therefore one will have beams which are wider at the butt than at the top. This superfluous width brings with it the advantage of more room for the butting together of decking if deck lengths, the full width of the bridge, are not available. One should balance out these variations in beam width by laying the beams top to tail alternately across the bridge during construction (see Diagram 48 on page 160).

Sawn beams should be quarter sawn and placed with the side nearest the centre of the log uppermost so that the natural bow of the beam will be upwards (see Diagram 42).

Beams are placed upon capsils which are placed upon piles. In the writer's experience, tropical hardwoods can often outlast the mild steel bolts used to keep them together. It is therefore his preference to design a bridge so that all the parts rest naturally one upon the other in compression and not to depend for strength upon bolts in tension. Thus, although capsils can be formed by bolting beams onto piles (Diagrams 46a and 46b on page 154) the preferred method is to make the capsil by trimming a log flat top and bottom, cutting a tongue on the top of the pile and fitting it into a slot in the capsil (Diagram 46c).

Once the piles have been driven and the tongues cut (Diagram 47 on page 155), the capsil can be cut. First trim the capsil smooth top and bottom, then mark in the positions of the tongues accurately to coincide with the pile tops. Allowing for the width of the space to be occupied by the beams plus about $^1/_2$ metre (20 inches) allowance either side (see bottom left of Diagram 41 on page 149), cut the capsil to length. The ends of the capsil should be bound with 8 or 10 gauge wire (see detail in Diagram 35) to prevent end splitting. The tongue slots can then be drilled out and trimmed to shape with chisels. The fit should not be too tight as this can cause difficulties when the capsil is lifted onto the piles. It can also induce splitting from the slots. Hot tar poured into the slots as soon as the capsil has been bedded onto the piles will help to preserve the wood and will fill the gaps between the tongues and the slots. If need be, a little sawdust can be used to thicken the tar to prevent it running through and being lost. Properly fitted in this way there is no need to use bolts or spikes to hold the capsil in place.

If bolt-on capsils (Diagrams 46a and 46b) are preferred, preparation of the pile is simplified and quarter sawn baulks (Diagram 42 on page 150) can be fitted in place, drilled and bolted home. It will probably still pay to tar the top of the pile at least.

Once the capsil has been fitted to the piles, cross braces can be added if necessary. They are normally just bolted into place as shown in Diagram 41.

Capsils can be fitted to concrete foundations as well as to piles, or even just laid in position on gabions, dry stone walling or prepared ground. It is the writer's preference always to use a wooden capsil between the beams and the bridge abutments because this gives the bridge span the integrity of a separate structure. This means that, in the event of a foundation cracking, crumbling or being undermined, the whole bridge span can be jacked up separately whilst the foundations are repaired. Furthermore, a span under load must flex slightly and the capsil becomes a cushion for such movement. Bearing this in mind it is advised that where rigid concrete foundations are cast with integral holding down bolts, the bolt holes in the capsil should be oval across the capsil, and the nuts not tightened down hard, to allow for the effect of the span flexing. The holding down bolts should be regarded mainly as a means of location and, should the span become immersed in an exceptional flood, as a means of preventing it from being lifted from its seating.

When capsils are to be seated on concrete or on gabions remember to leave space to get spanners onto the bolts that will hold down the beams. Equally, it will be necessary to get at the cast-in holding down bolts holding the capsil to the concrete between the beams. When using piles, the same problem will arise to the extent that beams cannot be placed directly on top of a pile but have to be spaced either side of, or between the piles in order to get at the bolts.

Once the capsils are in place the beams can be laid. If they are of round timber trimmed top and bottom they should be bound with wire as in the detail of Diagram 51 on page 163. However, if the trimming is to be done by axe and adze – note the position of the lashing in

DIAGRAM 48   Placing beams on capsils top to butt alternately to spread beam strength evenly

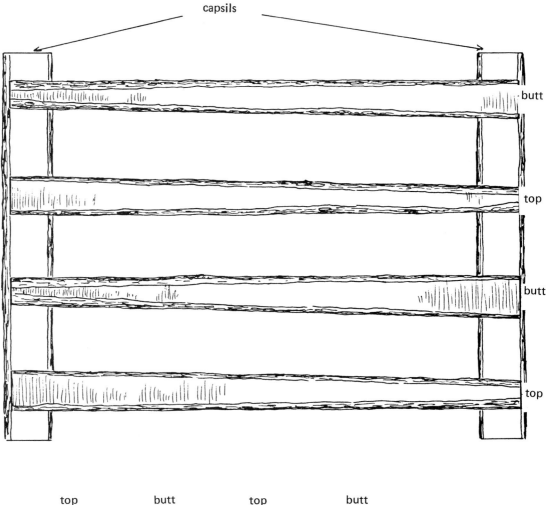

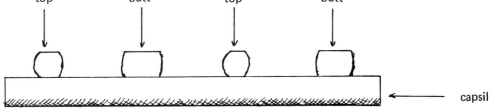

DIAGRAM 49   Concrete pier for wooden bridge

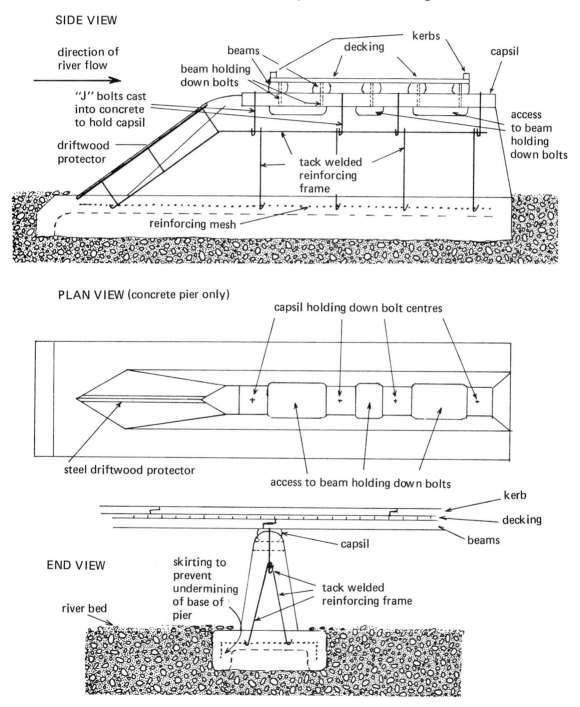

SIDE VIEW

direction of
river flow

beams

decking

kerbs

capsil

beam holding
down bolts

"J" bolts cast
into concrete
to hold capsil

driftwood
protector

access
to beam
holding
down bolts

tack welded
reinforcing
frame

reinforcing mesh

PLAN VIEW (concrete pier only)

capsil holding down bolt centres

steel driftwood protector

access to beam holding down bolts

kerb

decking

capsil

beams

END VIEW

skirting to
prevent
undermining
of base of
pier

tack welded
reinforcing frame

river bed

DIAGRAM 50   Bridge protector for wooden piled bridge

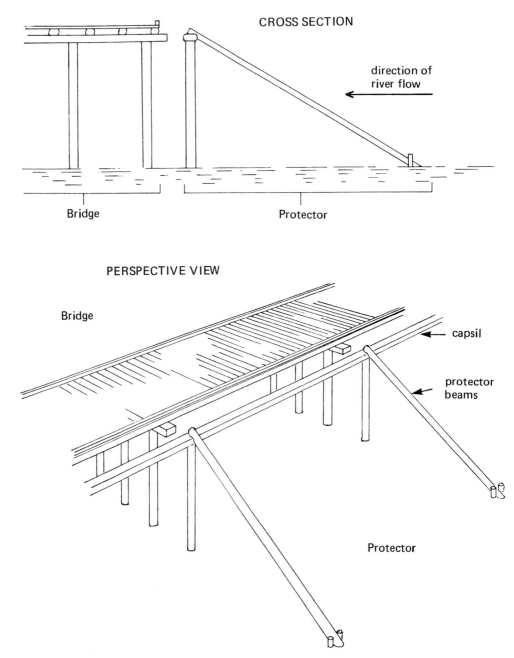

CROSS SECTION

direction of
river flow

Bridge

Protector

PERSPECTIVE VIEW

Bridge

capsil

protector
beams

Protector

DIAGRAM 51a   Binding beams with soft iron wire to prevent splitting

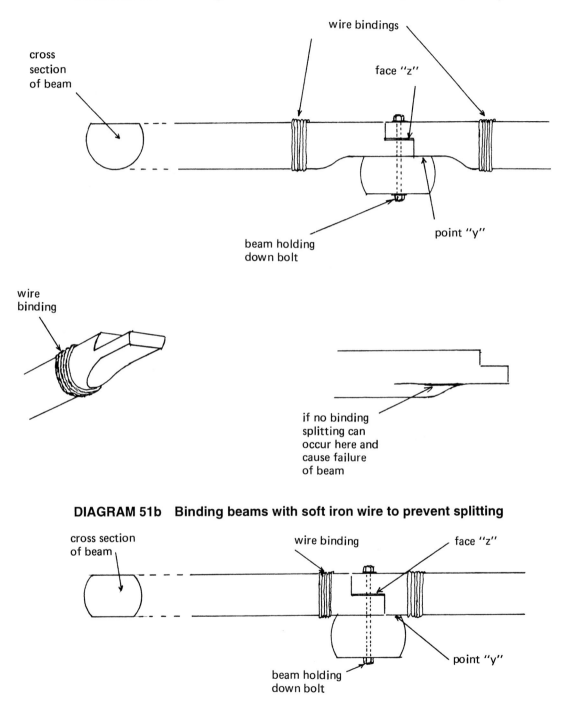

wire bindings

face "z"

cross
section
of beam

point "y"

beam holding
down bolt

wire
binding

if no binding
splitting can
occur here and
cause failure
of beam

DIAGRAM 51b   Binding beams with soft iron wire to prevent splitting

cross section
of beam

wire binding

face "z"

point "y"

beam holding
down bolt

Diagram 51a – it is not necessary to trim the bottom of the beam all the way along its length but it is necessary to bind it correctly to prevent failure due to splitting, as shown. If trimmed top and bottom in the sawmill then the binding should be tight up against the capsil as shown in Diagram 51b. Note also the method of scarfing the beams and bolting in place. It is essential that the right hand beam as shown in each case in Diagram 51 should rest firmly upon the capsil at point "y", if there is to be any tolerance gap it should be between the joint faces "z". If this is not observed, splitting of the beam on the right can start in from the top of face "z". The placing of hot tarred sacking where the beams rest on the capsil, and between the beams, is recommended and the scarf faces should be painted with hot tar or creosote. The bolts should be tightened only enough to locate the beams firmly, they should not be screwed down hard. The bolt holes should have tolerances of about 3 millimetres ($^{1}/_{8}$ inch) to allow for flexing of the beams under load.

Quarter sawn beams are likely to be deeper in cross-section than they are wide. They need not be scarfed on top of the capsil – except for the outermost beams – but can be laid side by side on one piece capsils (see Diagrams 46b and 46c) and butted end to end on double capsils (see Diagram 46a) as shown in Diagram 52. Because they are thinner than round log beams smaller diameter holding down bolts may have to be used. Remember to lay them bow upwards (see Diagram 42).

Do not allow the beams to be spaced further apart than about 60 centimetres (2 feet) edge to edge. This is to ensure that the decking does not flex under load otherwise the earth and gravel coverings will break down quickly in dry weather due to movement. On single lane bridges that will carry very heavy lorries, concentrate the beams under the runners or wheel tracks – neither the centre nor the edges of the bridge will carry much load. This too will help reduce flexing of the decking.

DIAGRAM 52   Laying sawn beams that are deeper than they are wide onto single and double capsils

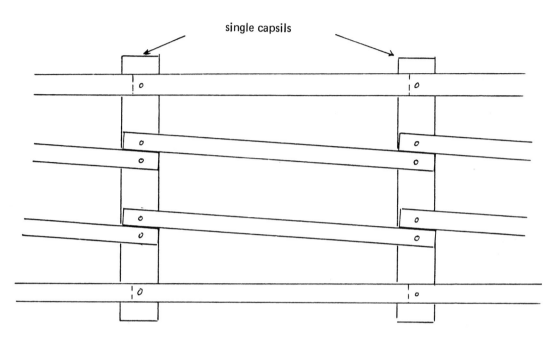

single capsils

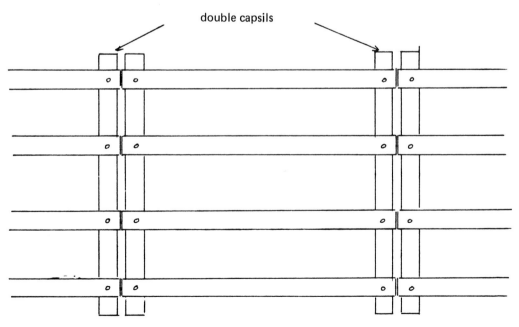

double capsils

DIAGRAM 53 Dovetailing of decking planks

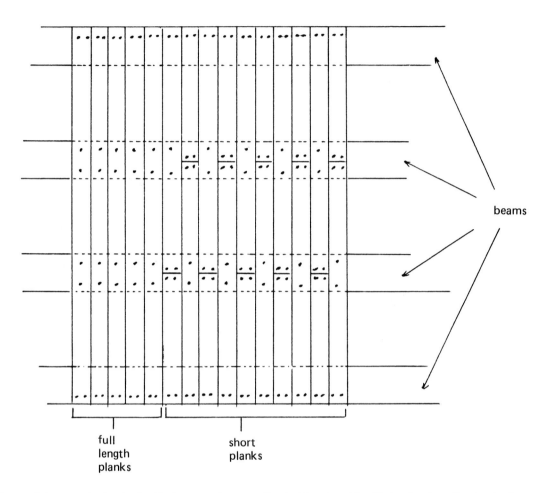

beams

full
length
planks

short
planks

If possible use deck planks that are as long as the bridge is wide. If shorter length planks have to be used alternate long and short planks to spread the strength evenly as shown above. Dark dots indicate deck spikes for nailing the planks down firmly.

DIAGRAM 54   Runners laid on decking to provide a smooth surface for vehicle wheels and to spread the load over the deck planks

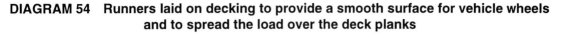

deck planks

kerb

runners

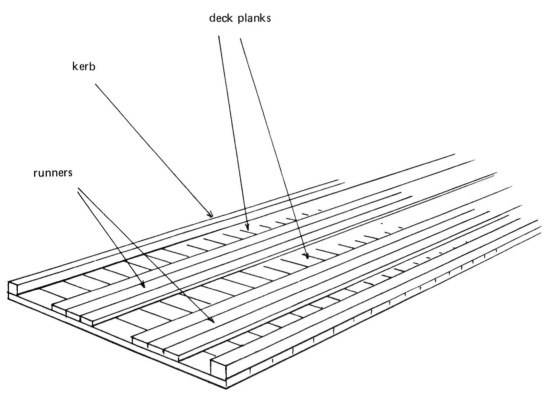

DIAGRAM 55a   Sawn timber kerb with bare decking and runners

DIAGRAM 55b   Round timber kerb with decking earthed over

holding bolt

runners

kerb

deck planks

beam

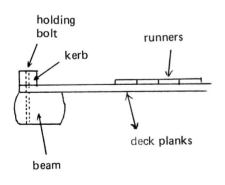

holding bolt

earth and gravel

kerb

deck planks

beam

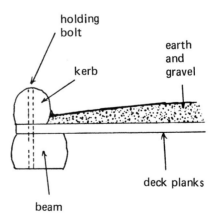
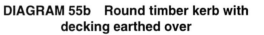

Decking, runners and kerbs

Where durable timber which can be cleft is available, and it is practical to earth over the bridges, then this is the best material to use. The cleft lengths can be cut anything up to the width of the bridge in length but, where lengths of less than that have to be used, they should be dovetailed in across the bridge as shown in Diagram 53 on page 166. The joints should be centred on the tops of beams to enable the ends of the planks to be properly spiked down where they abut – the same will apply to sawn decking. Where it is not practical to earth over the bridge, perhaps because it is too long or sufficiently durable timber is not available, sawn planks are the better material to use. After laying, they can be protected by the placing of runners spaced to provide tracks for the vehicles to run over (see Diagram 54 on page 167). The principal advantage of this is to spread the load of the vehicles across the planks and to stop the planks springing loose in time. When loose, the planks will rattle making a very consider-able noise as vehicles pass over them but this is only the most obvious sign of deterioration – the planks will rapidly crack, break loose and then shatter unless promptly attended to.

Cleft planks, which should average about 5 centimetres (2 inches) in thickness, can be laid in place by roughly matching the planks (see Pictures 5.7 & 5.8 on page 139), and then nailing them down to the beams with ordinary soft iron 6 inch nails. Almost certainly it will be necessary to drill holes for the nails through the planks and into the beams to a total depth of 4 inches; the remaining 2 inches will be drivable with a hammer in skilled hands (but an unskilled man is likely to bend all the heads over!). The same precautions may be necessary for the nailing of sawn planks too (if in doubt try a few first). If, however, the planks are not to be earthed over then nails are unlikely to be strong enough because the planks will suffer much more cruelly from the effects of direct contact with the traffic. It is then better to use square sectioned spikes, for which holes should be drilled through the planks but not into the beams, except perhaps just to get a start if the beam wood is very hard. The spikes are then driven in by sledgehammer. If runners are used to ameliorate traffic damage (see Diagram 54), they should be bolted to the planks using small coach bolts with washers or, if nailed, the nails should be clenched across the grain of the decking and their heads supported by washers or they will very quickly come loose.

When the decking planks have been laid the job can be finished neatly using a marking thread to put a line across the ends of the planks parallel to the edges of the two outermost beams but just outside them, and trimming off all the plank ends to that line. The kerbs can then be laid exactly to the plank edges and bolt holes drilled through kerb, planks and beams and the whole bolted neatly together (see Diagram 55a on page 167). Kerbs need to be about 10 centimetres by 10 centimetres (4 inches × 4 inches) at least to be effective. For a main road bridge 15 centimetres by 15 centimetres (6 inches × 6 inches) is recommended. They do not contribute to the strength of the bridge but they do stop loose planks from bouncing out of place and provide some measure of extra safety to prevent vehicles from running over the side. Where the bridge is earthed over, still larger kerbs are required (see Diagram 55b), both to act as kerbs and to prevent the gravel from falling off the bridge. For this round kerbs are better.

Handrails are not recommended, they are frequently damaged by machines or awkward loads crossing the bridge. If pedestrian traffic is heavy, it is better to construct a separate pedestrian walkway (on the downstream side of the bridge if flooding is a problem) to one side of the main bridge, extending the capsils for it if necessary.

Bridge protectors

Where rivers are liable to flood and carry substantial quantities of driftwood – whole trees sometimes – bridges with piers are at risk. This is especially true of bridges with piers or wooden piles supporting several spans. Most driftwood is swept down by a river in the early stages of a flood when the water level is still rising. As pieces of wood get snagged on the piles, they are unable to rise with the rising water level because of the pressure of the water holding them against the pile. The result is that as the water rises, more pieces of wood float over the first pieces to get snagged and in turn become lodged against the pile. This process continues with the result that a fence of driftwood is formed, held in place by the pressure of the water flow and the projecting end of the capsil above. The process is self generating to the extent that the more the flow is impeded the higher the water level upstream becomes and the higher the wall of driftwood grows until the bridge is swept away or overtopped. The fall of water over this wall of rubbish or indeed, over the whole bridge, can undermine the bridge very quickly by digging out the river bed between and immediately downstream of the bridge piers.

TABLE 4   Timber strength class factors

Strength Class	Factor	Approximate Average Density
1	25.0	1,060 kg/m^3
2	20.0	980 kg/m^3
3	16.0	830 kg/m^3
4	12.5	700 kg/m^3
5	10.0	600 kg/m^3

Note:
- Density is given because it has a very marked correlation with strength and therefore, in the absence of other information, is a good indicator that may be of help to the reader. But density is no indicator of durability: some very dense timbers rot easily and some relatively light timbers are durable.
- Normally only strength class 1 and 2 timbers would be considered for simple bridge construction. Strength class 3 timber, if durable, could be used if there is nothing better. All three higher classes could be considered for truss bridge construction if durable or suitably preserved.
- The factor is the bending strength parallel to grain in Newtons per millimetre squared obtained from tests of timbers within the strength class and for which a suitably conservative value has been derived for practical usage.

TABLE 5   Timbers by strength classes

Region	Name	Region	Name

Strength Class 1 Timbers

Region	Name	Region	Name
South and Central America	Balata # Greenheart §*# Guayacan # Manbarklak Massaranduba Morabukea Sucupira Tonka * Ucar	South East Asia	Balau *# Bangkirai *# Belian §*# Bitis Chengal *# Mesua Penaga Sal, heavy # Selangan Batu No 1 *#
Pacific	Buabua Hopea, heavy Hickory Ash Manilkara		

Strength Class 2 Timbers

Region	Name	Region	Name
Africa	African Oak Aganokwi Coula Difou Ekki §*# Elang Etimoe Limbali Missanda # Okan # Ovankol	Pacific	Green Satinheart Hopea, light Kempas Kwila *# Malas * Manilkara Pericopsis Wandoo * Yasiyasi
South and Central America	Acapu Aromata # Bagasse # Carreto Fustic Lapacho Mora	South East Asia	Binggas Giam *# Kelatong Kempas # Keranji §*# Merbau # Padauk

Region	Name	Region	Name
	Nargusta		Pyinkado
	Purpleheart #		Resak §*
	Tauraniro		Sal, light
	Wallaba #		Tualang

Strength Class 3 Timbers

Region	Name	Region	Name
Africa	Afrormosia §*#	Pacific	Karri
	Afzelia, yellow		
	&white §*#	South East	Apitong
	Esia §*	Asia	Balau, red
	Katibe		Eng
	Muhuhu		Gurjun
	Mutenye		Hora
	Odoko		Kapur *#
	Opepe §*#		Keruing #
	Sterculia, brown		Kulim
	Wildesering		Merawan #
			Punah
South and Central America	Angelin *#		Rengas
	Basralocus §*#		Selangan
	Courbaril *		Batu No 2 #
	Guaybo		Selangan,
	Kabukalli		red #
	Manni		Sundri
			Tembusu
			Yang

Notes:
§ = timbers resistant to teredo and therefore suitable for use in salt or brackish waters.
* = durable timbers suitable for use in fresh waters.
# = timbers suitable for use in construction work above water level.

Unmarked timbers are not necessarily unsuitable but there is not enough information available to the author that they can be so listed with assurance: the reader should check with the local forest department for details of the properties of timbers of interest.

Source: The tables above were assembled from information based both on the writer's experience and data provided by courtesy of the Timber Research and Development Association.

Where concrete piers support the spans, protection can be gained by sloping back the leading edge of the pier as shown in Diagram 49 on page 161. A steel plate or wooden fender on the leading edge may be necessary, both to absorb the impacts of the fast moving timber, and to provide a smooth face to enable it to float and slide up the protector and allow the free flow of water underneath it. The same principle applies to the wooden protector shown in Diagram 50 on page 162 and Picture 5.16 on page 143. In this case the protector is structurally separate from the bridge to prevent either the impact shocks or the weightload of the driftwood being transferred to the bridge. One protector beam per line of piles is usually adequate but additional beams between piles may be placed if it is felt necessary. The protector beams are secured in the river bed upstream both by being dug into the bed itself as far as is practical and by short piles driven into the bed to prevent them being forced aside. At the upper end the protector beams must lift the driftwood sufficiently high so that, in the event of the flood overtopping the bridge, the driftwood can float over and across the carriageway. The protector capsil must therefore be as high as the capsils of the bridge.

NOTE: after a flood has resulted in the deposition of quantities of driftwood on the protector, this must be cleared away immediately. If it is left it will both impede the rise of further driftwood up the protector when the next flood comes, reducing its effectiveness, and become a fire hazard to the bridge during the next drought.

It is usually only necessary to fit a protector to the uppermost bridge across a given river. By removing or cutting up the driftwood into small pieces at that point, all bridges downstream will be unaffected by the problem unless there are tributaries below that can discharge large amounts of driftwood. Further protectors may then be necessary.

The use of timber trusses

General

The coming shortage of long logs of durable tropical timber that have made bridge building of the type discussed earlier in the book so practical, will have to be countered by changes in the timbers and techniques used. There will be a trend therefore to developing ways in which smaller and perhaps less durable timbers can be used. A more sophisticated approach is unavoidable because of the need for jointing and preservation implied. Trusses will have to be constructed to replace the simple beams that were formerly used and, in the interim, it will pay to reserve the really durable long timbers remaining for piling and capsils. Eventually these items too will have to be made of other materials.

Truss bridges have to be built above any likely flood levels. They are less robust and less able to withstand being submerged and hammered by driftwood than the simple bridges made of solid beams. They are less heavy for a given strength and provide much more resistance to the flow of water across them. However, they will cost far less than the concrete or steel equivalent in most places and can give good service for many decades if they are looked after. Repair and replacement are also easier. Timber therefore, still remains the preferred raw material for most project road bridging requirements.

Bridges with the trusses above the decking

The writer has seen some very good timber truss bridges in the interior of Sulawesi that were built before the Second World War and were still sound thirty years later. These were of closed construction, that is, they had the trusses at the sides which rose above the decking and they were roofed on top to keep the whole bridge dry. They provided for single lane traffic and would permit a car or small lorry to pass through. Load height and width were thus restricted and there was evidence at the ends of damage sustained by impacts from vehicles that were not necessarily too heavy but certainly loaded too high or too wide. This type of construction therefore has its limitations both with regard to the size of vehicle able to pass through and, with the trusses giving their strength only at the bridge edges, with regard to the width of carriageway that it is practical to build for. If a two lane capacity is required it would have to be split into two separate carriageways with a third central truss between. Designing these sort of bridges will require the services of a qualified civil engineer.

Bridges with the trusses below the decking

Civil engineers at the Timber Research and Development Association (TRADA) of the United Kingdom have developed a design under contract to the United Nations Industrial Development Organisation with the trusses below the bridge decking. This method has the advantage of not restricting vehicle height and width with a superstructure, and of providing several trusses across the width of the bridge that can be spaced apart to support the decking as may be required (see Pictures 6.1 to 6.6 on pages 176 to 178). The decking can therefore be lighter than that required for the design mentioned above. Considerable attention has been given to simplifying and standardising the basic three metre long trusses so that they can be batch produced on a rugged home-made steel frame jig from sawn lumber. The trusses are secured with mild steel plates that can be manufactured in the average estate workshop. The bridge strength is related to the thickness and width of the lumber, the strength characteristics of the species of timber used and the number of trusses laid in parallel to support the decking. Since the basic trusses are standard, differing spans and strength requirements can be met by designing in more or fewer trusses. Tension is provided by mild steel rods and the individual trusses are connected together into assemblies on site and launched across the gap with simple cable and hoisting equipment.

The timber for the trusses does have to be accurately sawn to correct widths and thicknesses and be rigorously selected to have the grain running clear from end to end of each piece with no defects that could affect its structural strength. Accurate milling facilitates the simplicity of assembly of the individual trusses. Consistent timber quality is essential to enable performance to be guaranteed. TRADA bridges have been built and are in use in Kenya, Cameroons and in parts of Southern and Central America. In each case TRADA engineers have supervised staff training on the job and built the first bridge with each team under instruction. Thereafter the teams are left to carry on their work under the supervision of their own competent local managements. This approach to the art of wooden bridge building has much to commend it and it will undoubtedly become an acceptable answer to a problem that is really just beginning to be felt. (Project managements requiring further information are advised to contact TRADA directly, the address is noted in 'Further reading and useful addresses' on page 295.)

Alternatives to timber piles

Steel and reinforced concrete piles are likely to be too heavy for the home-made piling rig described in this book. One possible alternative is to use hollow steel piles and fill them with concrete after driving. The usual method of doing this is to cast concrete in the bottom one metre of a helical steel tube, pointing it with a former if required, and then to put in half a metre of dry cement mix as a cushion for the hammer. This is then driven to refusal, cut to height, a cage of reinforcing rods inserted and the tube filled with concrete, using a vibrator to help ensure a dense casting. Any cross bracing connecting lugs required would be connected to the reinforcing cage prior to casting. With a hammer that can be lowered into the tube this is a fairly simple operation but without such a hammer it may be necessary to obtain a strong wooden pole and insert it into the tube to use as a drift and drive that with a normal hammer. The pole would have to be a loose fit into the tube, cut square and very well bound at both ends to avoid splitting. After driving, the pole would be extracted and the tube filled with reinforced concrete. It is possible that large diameter toughened plastic piping could be used in the same way. Once driven, cut to height, filled and set, reinforced concrete under-capsils could then be cast onto the piles. One piece wooden upper capsils placed on them to form a bed for the trusses would provide an integral wooden structure, separable from the concrete for repair work.

Polewood bridges

A system has been developed (and patented by EMIR, see address on page 296) which uses polewood held in compression between steel connector units by steel cables under tension, for both bridges and large-span, low-cost buildings. It enables long spans to be put together without the need for permanent intermediate supports. It utilises what are often wasted or low value materials, requires minimal skill in the production of the component parts (all of which are easily manhandled) and their subsequent assembly. Equally, the structures can be taken apart for use elsewhere or for repair. The resultant bridges are flexible enough to stand moderate earth tremors.

The bridges can be designed with the roadways suspended from the structure or, as most projects will prefer, with the roadways supported above the structural members, to avoid the risk of accidental collision damage weakening the bridge.

The connector units consist of flat steel plates (see Pictures 6.7 to 6.11 on pages 179 to 181) with holes made in them to position the tensioning cables, and tubes welded to them to locate the poles. Whilst not contributing to the basic strength of the structure, they will hold it up until the cables are tensioned. They also serve to provide stability by resisting side loads resulting from violent gusts of wind, vehicle movements and, to some degree, ground movement. This is ment. This is important as it could well prevent a bridge from collapsing after sustaining some degree of foundation damage, until repairs can be carried out.

All the steel parts required for the building of a bridge could be manufactured within the average project's workshop, from materials readily available. Accuracy is required for cutting the poles to precisely the right length and to ensure that the ends are cut square. The ends of the poles must be turned to the correct diameter (machine-peeled poles treated with preservatives are ideal for this purpose) to give a close, but not a force, fit. Holes for rainwater to escape will be needed in the bases of the up-turned connectors, and the application of hot tar or bitumen will also help to prevent the ends of the poles rotting over time.

Tools required for assembly are a long jack (see Picture 6.9 on page 180) and temporary scaffolding, spanners for tensioning and holding-down bolts, etc.

The lifetime of bridges constructed in this way will depend upon the effective treatment of the poles with preservatives, preferably prior to assembly, the quality and cleanliness of the welds, effective drainage of the upwardly open tubes and the maintenance of regular protection to the steelwork.

Oil pipe bridges

The use of discarded oil pipes for bridge making is widespread in parts of South Sumatra, Indonesia, where oil companies have been active for many decades and it is not always economic for them to recover old pipe lines or, if they do recover them, they are not re-used but sold off as scrap metal. Under such circumstances, skilled welders are also likely to be available and possibly also the structural engineering knowledge to assist in the design of bridges made from these materials – the oil companies themselves often utilise old pipes for bridge building. Other areas in which oil has been drilled may offer similar opportunities for the construction of bridges from old pipework, at costs that will be far below the cost of bridges built from new steelwork or concrete. With a lifetime of two or three decades, perhaps more if well looked after, such bridges can be adequate to carry both project vehicles and road maintenance machinery, and thus be a cost-effective method of gaining access to land that would otherwise be too expensive to develop.

Examples of bridgework using old pipes are shown in Pictures 6.12 to 6.14 on pages 182 and 183.

Picture Group 6:   Timber truss, polewood and oil pipe bridges

6.1/6.2   Preparing the individual trusses (6.1), and launching a beam consisting of paired trusses, (6.2). Note the completed beam already in position in the background. (This series of photographs was made available by courtesy of the Timber Research and Development Association.)

6.3   Nailing down the decking onto the trusses. Note the extended pieces on the left that carry the handrail and the kerb being bolted into place on the right edge of the bridge.

6.4   The completed bridge. Note the runners placed above each pair of truss beams and positioned flush with the edge of the concrete sill. These minimise damage by grader operations and the "drop-on" effect of vehicles running off the road gravel and onto the wooden bridge runners and decking.

6.5   The completed bridge with the trusses well above anticipated flash flood levels.

6.6   Detail of the trusses from underneath the bridge. Note diagonal cross braces between trusses within each pair and horizontal bracing between pairs of trusses. The steel tension bars are infilled with blocks of wood and coated with rust resistant paint.

Polewood bridges

6.7/6.8 Polewood bridge connectors, straight and angled. Note that the plates intrude into the pipes, the accurately sawn ends of the poles abut immediately onto them. The small central holes in the plates are for the tensioning cables to pass through.

6.9 Jacking apart the plates in the span of part of a polewood bridge structure to enable the last set of six poles to be inserted into it. At this stage the structure is sufficiently "floppy" to enable the poles to be put in place (lower ones first, moving up to the upper ones as the jack is let down, little by little). The span is then allowed to fall back into position as the jack is let right down. Subsequently the tensioning cables (two in this case) will be inserted and, when tautened, the span will become rigid and immovable. Note the simple scaffolding required. None of the parts are too heavy for two men to manage comfortably.

As with the TRADA timber truss bridges, additional strength with the EMIR system is gained by adding more standard-sized poles and connector structures side by side until the required load-bearing capacity is reached.

6.10   Polewood bridge foundations are simple, reinforced concrete blocks with holding-down bolts embedded into them. In this picture the tensioning cable assembly has not be fitted.

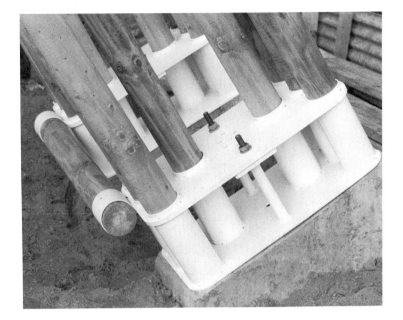

6.11   Polewood bridge foundation with tensioning cable fitted and tensioned. The four bolts in the centre are the tensioning bolts. Note that the cable has been greased and sleeved with plastic pipe to prevent corrosion.

6.12   An old oil pipe bridge in Sumatra. In this case, the oil pipe piers are topped off with scrap H-section girders and a wooden plank deck with runners. At the time this picture was taken it had already been in service for more than twenty years, according to the locals.

6.13   Close-up of the bridge above to show the structure of each pier.

6.14 Sheet steel runners on an old oil pipe bridge in which old oil pipes were even used for the decking. Not suitable where animals may have to cross, but effective and acceptable for vehicles and pedestrians – especially where the materials are virtually free of charge!

Picture Group 7:   Stone arch bridges

7.1   Stone arch bridge in sandstone with a supplementary arch of brickwork, three courses thick.

7.2   Detail of the dressed sandstone arch which forms the main strength member of a stone arch bridge over a canal. A crude gritty mortar was used between the stones to ensure contact over the whole surface of the dressed stones during construction. Both bridges are more than 250 years old.

7.3   Stone arch bridge on a steep mountainside road. Note the rocky river bed providing ideal foundations for this type of structure.

7.4   Detail of dry stone walling. The flat slatey stones are laid across the wall from outer surface to outer surface and act, in a sense, as tie bars. The walls have a marked batter, being almost a metre thick at the base and a little over half a metre thick at the top.

7.5   Stone arch bridge of dry stone construction several hundred years old. The height of the parapet wall is indicated by the vehicle on it. There is only a thin layer of river gravel between the arch and the road surface.

7.6   Detail of the arch stones and the way the parapet wall stones are fitted neatly into them. No mortar has been used in the arch but note how thin slivers of stone have been placed between the arch stones to wedge them apart at the top and ensure that they are truly aligned with the focus of the curve of the arch.

7.7   Detail of the brickwork on an arch bridge having a heavy batter on the spandrel wall.

7.8   Form work for a stone arch bridge – separate assembled sections of the form work prior to placing at the bridge site. Note the centre section (painted black) which can be dropped out before removing the side sections after construction of the arch. Any two sections will support a simple three piece additional sheet between them so that it is only necessary to construct three formers to support an arch five times the width of a single former. These additional sheets are not shown here.

7.9/7.10   Note that the arched part is supported on trestles and that the centre section can be dropped out – it is wedge-shaped – to facilitate the removal of the whole support structure. The supports are overlaid with steel plate which can be in turn overlaid with polythene sheet to prevent cement sticking to it. This will make the removal of the whole metal assembly easy. All the sections of metalwork are small and easily manhandled. The can be used many times.

Stone arch bridges (see Diagram 56 on page 190 for nomenclature of bridge parts; see Picture Group 7 on pages 184 to 188)

General

In Europe many stone arch bridges hundreds of years old are still used to this day, carrying loads of a magnitude far beyond the wildest dreams of their creators. They were built before today's sophisticated materials and engineering practices were available by men who were no more than skilled craftsmen. It may be felt therefore that in places where good sand and stone and good labour are plentiful, the stone arch bridge could be a viable alternative to the use of the hardwoods that are now becoming scarce in many places. However, it must be noted that of the bridges that still stand the majority are either founded on massive rock, on very firm ground which is not liable to scouring by the rivers that they cross or are built across rivers that flow sluggishly. Not many survive where less favourable conditions occur (and, as the stone they were built of was much sought after for other purposes, little evidence remains of the failures). This is not surprising – the essence of the arch bridge structure is that all loads are received and transferred in compression. There is no effective tensile strength in the materials used. Therefore, any settlement or failure in the foundations of the piers and abutments will lead to the collapse of the whole structure. This indeed is how the vast majority of arch bridge failures did, and still do, occur.

The need to provide sound foundations where good solid conditions do not exist naturally, has exercised the minds of bridge engineers for a very long period of time. The fact is that all the answers are pretty complex and expensive, and failures still occur. So there is no point in project management who are not civil engineers trying to do what is, for them, the impossible. If masonry bridges are attempted they should be restricted to places where the foundations will be on nearly ideal sites, and before making the attempt one would be well advised to consider the cost effectiveness of using multi-plate large-diameter steel culverting first. If this is out of the question then the notes that follow are intended to point out the problems, provide some guidance for the person responsible on site to decide whether this is an option or not, and enable a skilled mason on a project to have a go, albeit on a small scale under management supervision, to start with.

Problems with water flow

If a river is constricted in width it will attempt to maintain its flow rate, both by increasing the speed at which water passes through the constriction and (to the extent that increase in speed is an inadequate compensation) by gaining additional cross-sectional area for its flow in the vertical dimension until a new equilibrium is attained. It will rise in level and/or deepen its bed at the point of the constriction. In tropical situations, where floods are usually accompanied by considerable increases in the speed of the current, scouring of a susceptible river bed at a constricted point can be very deep and yet this depth is difficult to appreciate. This is because, whilst at full flood the river will sweep everything out of a deep sector of the bed, as soon as

DIAGRAM 56   Nomenclature of masonry bridge parts

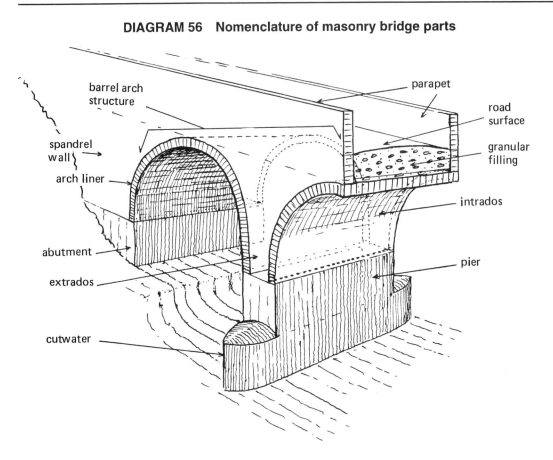

DIAGRAM 57   Pitting due to scouring at the foot of a square-faced pier

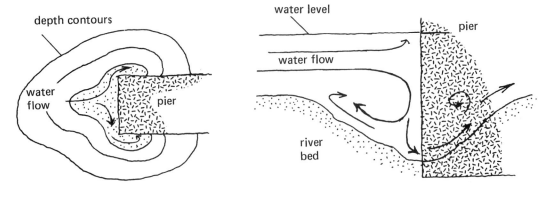

DIAGRAM 58   Pitting due to scouring downstream of piers

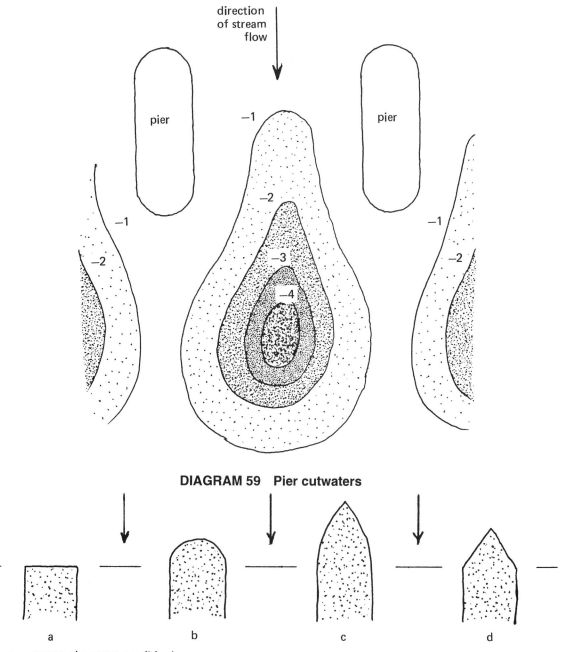

direction
of stream
flow

pier

pier

DIAGRAM 59   Pier cutwaters

a

b

c

d

a — square, the worst possible shape
b — semi circular, may still suffer slight pitting
c — gothic arch, good protection against pitting if parallel to stream flow
d — angled cutting edge, good where pier not quite parallel to stream flow, corners will shed laminar flow

the flow rate falls the hollow will be rapidly refilled again with gravel that is still being carried down from upstream. Thus, when the water is calm enough for human beings to inspect the situation, the evidence is totally obscured. The river bed looks much as it did before and there is no obvious reason to suspect that at the height of the flood the bed may have been very many metres deeper. This temporary deepening of a river bed in unconsolidated material can be two, three or even more times greater than the corresponding visible rise in water level at flood time compared with normal water level.

With wooden piled bridges of the type discussed earlier in this book, the piers occupy little space and – provided that the advice on adding spans rather than ramping up to the river edge with the aid of revetments and of providing bridge protectors where relevant, has been followed – they will not cause the catastrophic deepening of the river-bed described in the previous paragraph. Medieval stone bridges usually had piers that were between one half and one third of the width of the spans they supported. This was necessary to provide piers that were – bearing in mind that they had nothing better than crude mortar to work with – broad enough to be stable and robust enough to withstand damage from boats or floating driftwood. This meant immediately that the existing cross-sectional area below water level for the river to flow through was cut by 25% to 33% unless the river bed was allowed to widen by that amount, involving extra spans, to give a total cross-sectional area much as before. Indeed, most of these old bridges that are multi-spanned show just such a situation unless the river bed is on solid rock and too steep to flood to any significant depth.

Even when this adequate provision has been made, water flow will still raise problems where the river bed is not hard enough to resist erosion by scouring. Pits can form at the foot of a pier on the upstream side (see Diagram 57 on page 190) and also downstream (see Diagram 58 on page 191) between piers. To some extent the upstream pitting can be reduced by rounding off or sharpening the leading edge of the pier (see Diagram 59 on page 191) and by sloping the leading edge away from the flow of the water as shown in Diagram 44 on page 161. If the pier is not in perfect alignment with the water flow these effects can be exacerbated. Both forms of pitting can be ameliorated by the placing in the pits of stones too heavy for the river to shift, but this is not something that can be done once and forgotten about. Periodic inspection and topping up will be necessary. The risk is that, if neglected, the pits can in time become big and deep enough to undermine the foundations of the piers or abutments. It is therefore recommended that PROJECTS SHOULD NOT ATTEMPT TO BUILD MASONRY ARCH BRIDGES UNLESS REALLY GOOD FOUNDATION CONDITIONS EXIST AT THE BRIDGE SITES.

Abutments and piers

For single arch bridges over ravines with hard ground either side, the abutment foundations should be dug into the rock sufficiently deeply to ensure that they will be secured and immovable (jackhammers, drills and even explosives may be required). They should be set at right angles to the thrust of the arch (Diagram 60 on page 193). It is assumed that in such a position they will not be liable to any scouring from the river.

DIAGRAM 60   Abutments set in rock sides of ravine at 90 degrees to thrust line of arch

see also Picture 7.3 on page 185

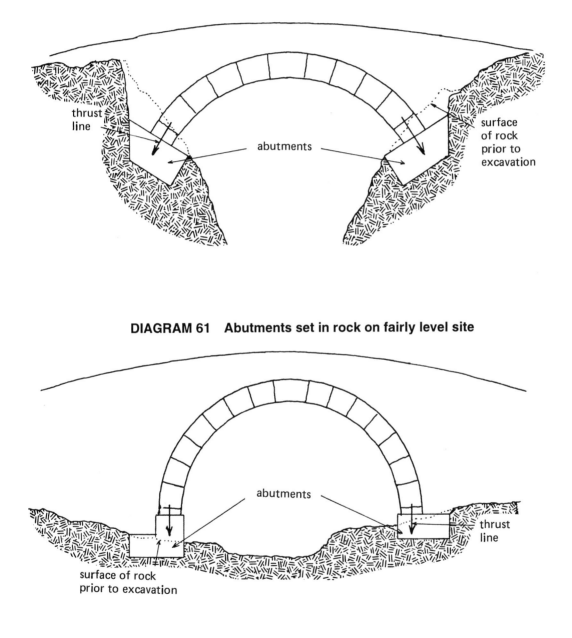

DIAGRAM 61   Abutments set in rock on fairly level site

For single arch bridges on a relatively flat site (Diagram 61 on page 193) the situation of the abutments is similar to that of piers in that the thrust is taken vertically and therefore differences in foundation level are unimportant. What is important is that the foundation is firmly secured on hard massive rock that is not susceptible to scouring by the river. The abutments should not be allowed to project into the river in such a way as to impede or deflect its normal flow.

It is assumed that no project will consider the construction of bridges of more than two or three spans without having qualified engineering assistance but, if piers have to be built, then they have to be founded with as much care as the abutments and shaped to reduce pitting and, if liable to flooding, to cope with driftwood. Methods of construction will depend upon what materials and what local skills are available. If the project can secure the services of good stone masons and there is either good stratified (lamellar) stone, or locally made brick of good quality, the abutments and piers can be built up in much the same way as is used for building massive walls. Alternatively, formwork of wood can be made and the stones placed against the inside of the formwork whilst cement is constantly poured in behind to fill the centre of the block, using metal reinforcement if required. A vibrator will probably be needed to ensure that these basic blocks of the bridge structure are strong and dense. If only irregularly shaped stone is available it is best used to make good reinforced concrete and cast the whole of the bridge structure, abutments, piers and arches, in stages. To be able to do this economically it has to be assumed that the project will be running a quarry and crusher which would supply both crushed and uncrushed stone to site. Buying in all the crushed stone required would indicate such a major expenditure in an unfamiliar field that it would probably be better to put the whole operation out to contract to civil engineering contractors. If the pier is cast in reinforced concrete, this will enable a much less massive form of pier to be built, perhaps down to one eighth of the span in thickness, compared with the one half to one third used in the medieval masonry bridges.

Arches

The cost of the formwork for an arch is likely to be similar to or greater than the cost of the materials for the arch itself. This point must be made because, to someone not used to bridge building, this may well come as a surprise and because any attempt to skimp on the strength of the formwork can be very dangerous: the arch will be unable to support itself until the arc is complete. It is possible to construct the bridge as a series of narrow arches side by side with interlinking reinforcement until the full width of the bridge is completed. This method is more suited to working in concrete than stone and will slow down the work but, if the time is available, it can save on formwork costs provided that the formwork has been designed to be reusable. If spans can be standardised then the formwork could also be used for more than one bridge span and costs can be usefully spread in this way also.

If stratified stone can be used it is best laid with a cement mortar containing around 500 to 600 kilograms (1,100 to 1,300 pounds) of cement per cubic metre (1.3 cubic yards) of sand and, as shown in Diagram 62 on page 195, not less than two courses. This way of laying the stone is immensely strong and economical in the use of cement. (See Pictures 7.1 to 7.6 on

DIAGRAM 62   Laying and interleaving courses of flat stones in the barrel arch structure of a masonry arch bridge

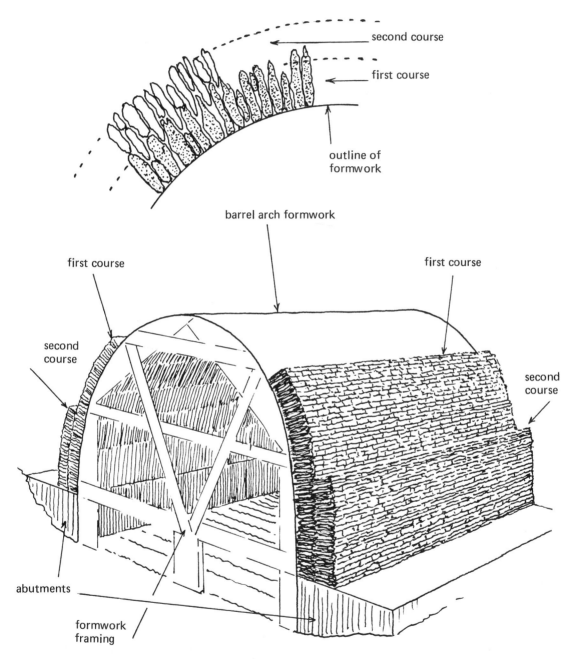

second course

first course

outline of formwork

barrel arch formwork

first course

first course

second course

second course

abutments

formwork framing

The work of building up the stone work should proceed evenly on either side of the formwork in order to maintain a balance of stresses acting on it.

DIAGRAM 63   Formwork for casting arches: plates of permanent material forming part of the shuttering between sections being cast

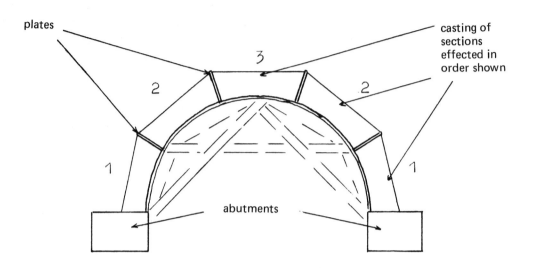

DIAGRAM 64   Sandbox for releasing formwork

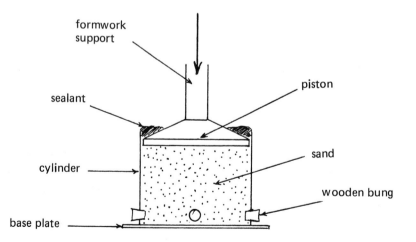

pages 184 to 186 of stone bridges laid without the benefit of any cement at all that have been in use for several hundred years.) Laying brick is done similarly but, as the brick will not interleave the way the stone can, rough surfaced bricks should be used and every effort made to get a really strong bond with the mortar. If concrete is being poured to make the arch, then the formwork should include plates of precast concrete or other permanent material that can be left in the finished bridge as shown in Diagram 63 on page 196. This will enable casting to proceed block by block over the arch whilst avoiding the problems of having large quantities of wet concrete to control on an inclined surface. The casting would be the full thickness between the intrados and the extrados. Alternatively the blocks could be pre-cast accurately on the ground, lifted into place and subsequently cemented together. In all these methods, work should proceed equally from the pier top or the abutments on either side towards the highest point of the arch, to ensure that an even balance of loading is maintained on the formwork.

Spandrel wall

Once the arch is complete, work on the spandrel wall can commence. It is preferable that work should proceed evenly from all four corners of the arch upwards to the top of the arch and on into the parapet. A slight inward batter, perhaps 1 in 50 from the vertical, is often given to the outer face of these walls which are usually made substantially thicker at the bottom than at the top. The rock and earth filling can follow the wall work upwards. Tie rods may be used at intervals and laid into the walls as they are built up of stone or brick, or are cast. The filling material should be granular and free draining. Clays that can flow under pressure over time or which will expand and contract between wet and dry seasons, are not suitable. The filling should be not less than one metre thick over the top of the arch and preferably more. The filling itself will provide a bridging effect, both cushioning and spreading the loads inflicted on the arch by vehicle wheels as they pass over it. The filling should be shaped and cambered to shed surface water off the bridge. This means that in a two span bridge the highest point of the filling would be midway between the two arches. For longer bridges it will be necessary to put in drainage outlets through the walls at appropriate intervals.

Removal of formwork

One of the simplest ways of removing formwork is to use the sandbox method (see Diagram 64 on page 196). This involves mounting the forms right at the very beginning on pistons of wood or thick metal plate set in drums filled beforehand with sand. The sand should be absolutely dry and the whole contraption protected to keep it so. In the drums, well below the level of the pistons, wooden bungs should be fitted into two or three holes around the circumference of the drums so that they can be removed from the outside. These remain in position until the formwork is due to be removed. The formwork can be extracted in stages: the bungs can be removed simultaneously from each drum to let the formwork down, measuring limited amounts of sand out of each hole, and then closing the bungs until all drums have lost equal volumes of sand when the next release can begin. This will ensure no uneven strain being put on either bridge or formwork.

This method will not suit all types of formwork and other means may have to be devised. Jacks, opposed wedges and the like can be used. It is however, better to use a means of gradual release rather than one which results in the sudden collapse of the formwork – not that the arch is likely to collapse but, if formwork is to be used several times, it has to be treated with respect.

Notes:

- It was interesting to observe that the arches of a three hundred year old masonry bridge recently opened up for repairs in Sussex, and examined by the writer, were found to be only 33 centimetres (13 inches) thick. The arches were perfectly sound. The repairs were being effected to the spandrel walls, but the opportunity was taken to drill several small holes through the arches in order to facilitate better drainage of the chalky infill which had become damp and had spread under the weight of modern main road traffic. This had caused the spandrel walls to bulge and crack. Proof – if it be required – of the immense strength of the stone arch. A warning too, that infill material must be selected with care. Following completion of the spandrel wall repairs and refilling with a more suitable infill, the bridge was reopened to normal traffic.

- The method by which the so-called "cave" houses of the people of the upper Yellow River in China are constructed is also interesting. The houses are built below ground level by digging deep and narrow trenches into the side of a valley, preferably just below the level surface of the loess soil that forms the surrounding high plains. These trenches form the moulds for the house walls which are then built within them of soft dressed stone or cast from stone and mortar. Some of the soil removed is heaped in a mound on top of the undisturbed soil between the walls and shaped to become the formwork for the dressed stone arch ceilings which are then laid. A slurry of mortar is poured over the stone arch and allowed to set (the effect being to produce an arch very much like that illustrated in Picture 7.2 on page 184) and a low spandrel wall is built along the front of the house. The soil from between the house walls and beneath the stone arches is then excavated (in so doing the rooms of the house are formed) and a thick layer of it is placed over the arches, between the arches and behind the spandrel wall, to provide a flat open courtyard on top of the house. In an area of low rainfall, but subjected to scorching heat in summer and searing frosts in winter, this method of construction provides a durable dwelling that is well insulated against these natural extremes of temperature.

  The method of construction has relevance to the construction of stone arch bridges in an alluvial area where the bridge can be built to one side of the existing river course. Advantage can then be taken to dig down well below the level of the riverbed during the dry season to construct the pier and abutment foundations. The soil excavated can be heaped between the foundations and used to support or become part of the formwork, for the arches subsequently. Construction of the bridge can then be effected on "dry land" followed by the excavation of the spans between the arches and the re-alignment of the riverbed to flow through them.

This method of construction would avoid all the problems involved in building the formwork and of lifting all the stone onto it to make the barrel arch structure of the bridge.

Fords and Irish bridges (see Picture Group 8, pages 208 to 211)

Simple fords

Braided rivers which carry large quantities of gravel and stones, usually at times of flash flooding, present special difficulties. They frequently change course so that a bridge built over the presently existing channel may at any time suddenly be left high and dry, becoming superfluous. In such unstable situations, deposition is occurring and the land is being steadily built up with the proceeds of erosion from the hills and mountains above. Only massive bridges which dwarf the river and can accept changes of course and deposition are going to be fully effective. Such structures are not relevant to this book, as their costs would be beyond the means of any normal agricultural or forestry project to fund. Fording such a river is usually practical but one has to accept that during times of flood it will become impassable until normal flow returns, usually only a matter of a few hours. At that point in time, it will probably be necessary to smooth out the roadway where the flood waters have passed over it, and the ford through the channel where the river flows, to enable the smaller vehicles to negotiate the crossing. Between floods, fords will still need frequent inspection and perhaps further maintenance, to ensure that vehicles can always pass safely through them.

For gravels up to about fist size, a grader can normally cope unaided, provided that it has a front-end dozer blade to push down the banks which form on the outer curves of the river (see Picture 8.2 on page 208), prior to using the main blade. This will protect the turntable, and/or actuating rams, from harm. Where the stones are substantially larger – rounded rocks and boulders – or where whole trees have been swept onto the ford site, it will probably be necessary to use a front-end loader or a bulldozer to clear and level out the path, as the use of a grader may not be practical. In such situations, the tracks of a bulldozer will probably be the most effective tool for smoothing the bed sufficiently for vehicles to use it without harm. The dozer should clear at least one-and-a-half times its own blade width to enable the tracks to crush and flatten the stones in the middle of the road, and to prevent protruding rocks between wheel tracks doing damage to the engine sumps of smaller vehicles. If a self-propelled vibrating steel roller or a cage roller is available, use it to finish off the work.

Floored fords and Irish bridges

In other situations, concrete-floored fords and Irish bridges are an acceptable low cost way of getting across a river, if only sand, gravel and cement are readily and cheaply available. However, their use is limited to situations in which the river beds are hard.

Site selection

Essentially, the site should be located on a firm sector in the river's bed, even if it is rocky – rocks big enough to be a nuisance can be blasted out of the way and the resultant stone used during construction.

Preferably, the site should be found at a point at which the river's course is both relatively straight and is confined between banks which are slightly higher than the surrounding ground, so that they are not liable to erosion. Such a site will almost certainly be regarded as an obligatory point during the layout of the road system of which it will form a part.

When the site has been chosen, examine it for any indications of the flood levels to which the river can reach, for example floating debris caught and left hanging in trees and bushes, "tide" marks on the banks, and so on. With the flood levels established, measure the cross-sectional area of the channel in flood and use this information in designing the structure of the proposed crossing.

Design and construction considerations

Both fords and Irish bridges are a shallow U-shape to allow, but contain, the passage of flood water (see Picture 8.3 on page 209 and Diagrams 65 and 66 on pages 203 & 204).

Care should be taken to ensure that concrete-floored fords do not raise the level of the river bed and constitute a bar across the river. If they are proud of the bed, the water flow will speed up as it passes over the ford and will then drop off it and could begin to excavate the bed immediately downstream of it. This would lead to undermining and the collapse of the ford floor. A ford is essentially part of the river bed and will always be wet unless the river itself stops flowing.

A structure which is raised above normal flow, and which accommodates normal flow in one or more conduits built within it, is usually referred to as an Irish bridge. It has the advantages of:

- being cheaper to build than a bridge high enough and strong enough to be clear of all anticipated flood levels;
- not requiring the vehicle to go so far down into the river bed and to climb out again, as it would have to do with a ford – important for long vehicles, lorries with trailers and so on, which do not negotiate such obstacles easily;
- enabling most vehicles to maintain something near their usual road speed;
- enabling vehicles to "ford" the river on the bridge at times of low flood levels (for instance in a situation when a normal ford would be too deep to be passable); and
- enabling cyclists and pedestrians to cross dry-footed in normal flow conditions.

The shallow U-shape profile of an Irish bridge should result in a channel for water flow that is at least equal in cross-sectional area to that of the untouched river bed at full flood. It is wise to allow for it to be some 50 per cent greater and, in this calculation, to neglect the cross-sectional area of the conduits which may become blocked with debris during the first few minutes of a flash-flood. This implies the possible need to widen the river bed for a short distance, both above and below the bridge site. This will ensure the smooth and uninterrupted flow of the river at full flood. Turbulent flow can lead to erosion where it is not wanted.

The whole run of the vertical profile of the bridge should be a gentle curve (NOT a V-shape, since this will cause vehicles to pound the structure at the point of the V, where it is thinnest), with the ends rounded off to facilitate grading operations on the gravel roadway at each end, without risk to the grader's blade.

Whilst most people place the conduit pipe or pipes centrally (see Picture 8.3 and Diagram 65), this is the point at which the structure is at its thinnest. It is, therefore better, if the shape of the river bed permits, to space two or more conduits either side of the centre where the formation is thicker (see Diagram 66). This allows the accommodation of pipes of a larger diameter at a point where the structure is naturally stronger.

It is essential that adequate protection is given to the outfalls of any culvert pipes and to the downstream side of an Irish bridge, to prevent undermining (see Diagrams 67a and 67b on page 205 and Diagram 68 on page 206). The upstream side is not usually at risk provided that the foundations are set in deep enough. However, the downstream face will assume the character of a waterfall during times of flooding and the destructive power of the water dropping onto the river bed can be considerable. (See Pictures 8.4 & 8.5 on page 210.) Hence the need for an apron that will deflect the direction of flow through 90 degrees from vertical to horizontal, and slow the laminar underflow before it moves onto the natural river bed.

The radius of the deflecting arc should be related to the maximum depth of water that will flow over it, and be at least equal to it (preferably up to double). Then add a similar distance for the flat part of the apron containing the embedded stone, in order to break the force of the laminar underflow (see Diagram 67a). This rule of thumb applies where the depth of water at peak flood flows is not more than the height of the road surface above the river bed. Where the depth of water during floods is markedly greater, the radius of the arc is less important than the extent of the apron of level concrete. This should be long enough to enable any downward force derived from the waterfall effect to be completely absorbed and levelled off with the bed of the stream (see Diagram 67b). This could be as much as the width of the road itself. It is better to spend a bit more than is necessary making an apron that is too wide, rather than one which is not adequate, resulting in the structure becoming undermined over the years.

If the river bed is gravelly rather than rocky, and water velocities are high during floods, there may be some risk of undermining on the upstream face. Deeper foundations are one answer, but the problem can also be ameliorated by sloping the retaining wall to facilitate the flow of the water up over the bridge, thereby reducing the turbulence where the flow meets the wall, which would lead to undermining (see Diagram 68).

It is not a good idea to put kerbs on an Irish bridge. They can trap debris, particularly tree branches and roots, which would otherwise flow across unimpeded. They can also trap mud and make the bridge dangerously slippery after flooding. For similar reasons, the use of marker posts along the run of the bridge, which can trap floating branches, bamboo and the like, is also not recommended. If the bridge is commonly forded during low floods, it is better to use markers on the upstream side only, shaped somewhat like bridge protectors (see Diagram 69 on page 207). These should be high enough to show above the water surface when it is safe to ford, with the clear implication that, if they are not visible, vehicles should not attempt to cross.

Rock fills laid down as foundations should be dressed off with stones and thoroughly compacted to ensure that there will be no settlement under the concrete running surface after it has been laid. It may be advisable to work a dry cement mix into this prior to compaction, knowing that in time it will moisten and set, to form a very durable foundation. Retaining walls should be anchored a good half metre or more into firm ground, or founded on hard bedrock.

There are arguments for and against the use of steel reinforcement:

- it will strengthen the concrete and spread the loads placed upon it;
- it will help hold in the retaining walls against spreading (equivalent to the "spandrel" walls of a stone arch bridge – see nomenclature, Diagram 56).
- however, if the reinforcement rusts, it can also swell and contribute to the break-up of the surface and make repairs that much more difficult, so avoid the use of it, if possible, under saline conditions.

If there is no risk of foundation movement or undermining, reinforcement is probably unnecessary. Conversely, if there is risk of settlement, it is probably essential to hold the structure together. The need to build up upon it in a decade or two's time, if settlement has occurred, will then have to be accepted, using the original as no more than a foundation layer. Further movement thereafter is unlikely.

Use a strong structural concrete mix for the whole operation, as both fords and Irish bridges are subject to heavy loads and strong currents.

Always try to carry out work during the dry season, if there is one, in order to avoid the risk of being overwhelmed by a flood before the construction is completed, and to minimise the problems involved in diverting the river during the process. If the river does not cease flowing, it may be necessary to divert it around the site in a temporary channel until the work has gone far enough to enable it to be allowed through the conduits. If possible, try not to have to make joins across the main structure of the bridge. This is not always achievable and the bridge may have to be built in two halves, another good reason for placing the conduits away from, and either side of, the centre of the bridge as shown in Diagram 66 on page 204.

DIAGRAM 65   Irish bridges with centrally placed culvert pipes

LONGITUDINAL CROSS-SECTION

round-off ends to reduce risk of damage to, or from, grader blade during maintenance work

gently rounded shape to prevent 'bottoming-out' stresses to lorry or bridge

culvert or culverts nestled in gravel

CROSS-SECTION ACROSS WIDTH OF BRIDGE

downstream side

upstream side

apron to protect against undermining

protect against driftwood if necessary

kerbs are not necessary – use corner posts

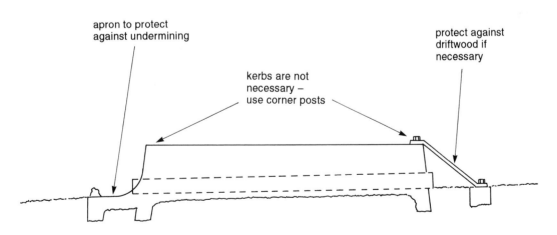

DIAGRAM 66   Irish bridges with culvert pipes placed at the river's edge

Where the stream bed is wide and flat, bigger culvert capacity can be
gained without making the Irish bridge too high by placing the culvert pipes
at the edges of the stream. The lowest point of the bridge can be level with
the tops of the culvert pipes affording immediate relief as soon as they
are flowing at full capacity. (Vertical scale exaggerated.)

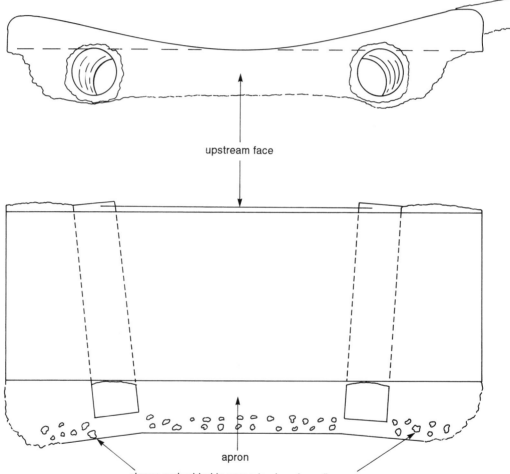

upstream face

apron

stones embedded in apron to slow down flow rate

Note that culvert pipes are laid to give convergent flow to reduce risk of
erosion at stream edges.

DIAGRAM 67a Protecting the downstream face of an Irish bridge subject to moderate flood depths

In this situation the apron does not need to be very wide provided a smooth transition from vertical to horizontal flow is achieved. Stones embedded into the toe of the apron will retard flow rates and protect a gravelly river bed from erosion. These precautions are not required if the bridge is founded on massive rock.

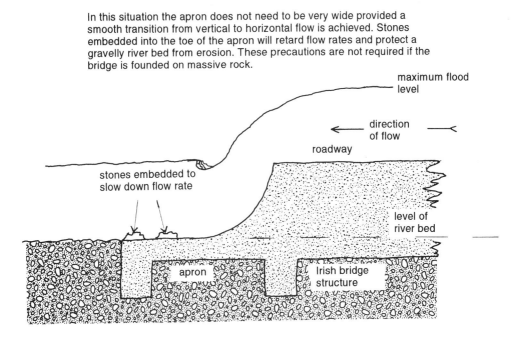

DIAGRAM 67b Protecting the downstream face of an Irish bridge liable to very deep floods

In this situation, velocities are likely to be much higher and a very much wider apron is required to ensure non-turbulent flow. Stones embedded in the far end of the apron will slow down laminar flow to reduce the risk of eroding a gravelly river bed. These precautions are not necessary if the bridge is founded on massive rock.

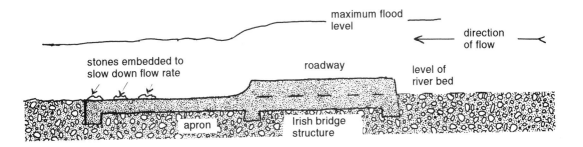

DIAGRAM 68   Protecting the upstream face of an Irish bridge against undermining

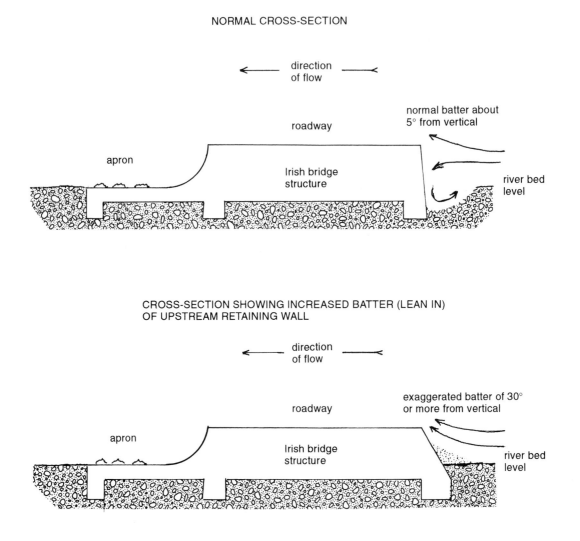

NORMAL CROSS-SECTION

direction
of flow

roadway

normal batter about
5° from vertical

apron

Irish bridge
structure

river bed
level

CROSS-SECTION SHOWING INCREASED BATTER (LEAN IN)
OF UPSTREAM RETAINING WALL

direction
of flow

roadway

exaggerated batter of 30°
or more from vertical

apron

Irish bridge
structure

river bed
level

A retaining wall with a batter of some 30° is quite likely to lead to a little
deposition of coarse sand or fine gravel against the upstream face
whereas turbulence will tend to undermine a vertical or normally battered
retaining wall. These problems do not arise if the Irish bridge is founded on
massive rock.

DIAGRAM 69   Markers for fording a submerged Irish bridge

CROSS-SECTION

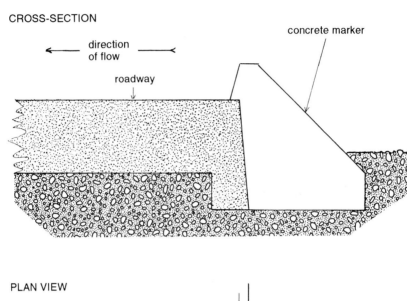

PLAN VIEW

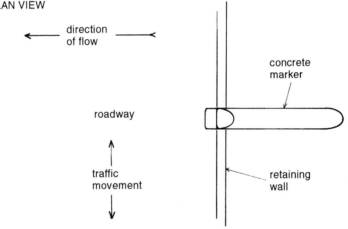

The height of the tops of the markers above the roadway should be the maximum depth at which vehicles can safely ford the bridge during flood times.

The markers are shaped to allow driftwood to move up and eventually float over the bridge as the water level rises beyond the safe fordable limit. They should preferably be cast integrally with the rest of the structure, have rounded upstream edges to minimise impact damage and be spaced not more than two to three metres apart.

Picture Group 8:   Fords and Irish bridges

8.1/8.2   Fords on braided rivers can spring nasty surprises when in flood and the water is no longer transparent (8.1). Someone should always walk the ford with a stick to test the depth before a vehicle enters it, otherwise one may drop over the edge of a sill into a newly formed deeper channel, such as that shown in 8.2, with disastrous results. It is much easier for the operator to level-off the ford with a grader after the flood has gone, and less likely to result in damage to the machine, if a front-mounted dozer blade is fitted. (Photographs courtesy of Simon Le Gassicke.)

8.3   A small Irish bridge in the bed of a stream liable to flash flooding, it lays in an area of shallow-soiled grasslands on level gravels between two areas of Eucalypt plantations. The dip in the centre is gentle but the ends would have been better extended a little further.

The downstream face, shown, lacks an adequate protective apron. This would have had no ill effects if the stream bed had been of solid, massive rock. However, it is on gravel and so has been slightly undermined below the normal water level for about two metres either side of the outfall pipe. Without attention, this will allow the "spandrel wall" nearest the foreground to break away and fall into the stream bed in the future.

In this situation, further undermining can be prevented by the construction of an apron in the dry season, as shown on the left side of the cross-section at the bottom of Diagram 65 on page 203. The apron should be below the normal wet season water level and stones set into the apron to reduce flow speed are desirable, as shown at the foot of Diagram 67a on page 205. A later addition like this will never be as good as doing the job properly in the first place, because the join will not be as strong as a one-piece integral casting, but it is better than leaving the bridge to deteriorate over time.

8.4   A large Irish bridge built with no protective apron on the downstream face. The central square-section concrete culvert has been undermined, the floor has collapsed (8.5, opposite) and the foundations of the side walls are already substantially undermined. The whole of the central sector of the structure is beginning to crumble and subside, the evidence being visible on the roadway (8.4). Attempts have been made to shore up the "spandrel walls" with gabions on the downstream face, but these too are being undercut (middle right of 8.5)

8.5   The repair of this river-crossing will now be an expensive operation involving the removal of much of the existing bridge before work can begin. During the period of the repairs the road will, in this case, be completely impassable.

RIVER CONTROL

General principles

The power of moving water

Stream or river control may become necessary in order to protect bridges, culverts or even road formations where rivers may flow close to them. Before considering how this work should be tackled, some elementary understanding of the "enemy" will help.

The ability of water to carry detritus along with it is related to:

- The volume and velocity of the flow of the river.
- The specific gravity, the size and shape of the fragments of rock carried.

This capability increases as to the sixth power of its increase in velocity. For example, a doubling of flow rate implies an increase in carrying ability of sixty-four (2^6) times. The specific gravity of water is stated sufficiently accurately for the purposes of this argument as 1. The specific gravity of most rocks falls between 2 and 3. Therefore, the effect of immersion in water is to reduce by between one half to one third, the effective weight of the rock. Under normal riverine conditions at a flow of 2 kilometres per hour gravel of up to 2.5 centimetres is moved. At 3.2 kilometres per hour gravel of 6 to 7 centimetres diameter is moved. It is easy to appreciate that, at full flood, with speeds of anything up to five to ten times the flow rates exampled above, big rivers can and do roll boulders weighing hundreds of kilograms down their beds.

The phases of river flow

Streams and rivers have three completely different natural phases in their courses. These are:

- The torrent phase.
- The floodplain phase.
- The estuarine phase.

The torrent phase comprises those sections of a river where the water is always fast flowing down a steep gradient, usually in a V-sectioned valley, and where erosion occurs mainly in the river bed itself. As this is deepened to the point at which the banks become unstable, they may slide into the river to be added to its load of silt, sand and gravel.

The floodplain phase applies to that part of a river's course in which the fall of a river is very slight and in which, as a result, flow speeds are markedly slower than in the torrent phase. The formation of a floodplain is caused by a barrier to the river's downward descent; this barrier may be the sea, a fiord or lake, or a large rock sill resistant to erosion. Whatever the cause, the river will be depositing material as a direct result of the loss of flow speed upon leaving the torrent stage. Following heavy rains in the river's headwaters the floodplain can become flooded (hence its name) and this usually results in the deposition of all the coarse material near the foot of the torrent course and progressively finer materials, silts and muds, further away from it over the areas inundated.

The estuarine phase begins when the river enters a sea or a lake. There the flow will be affected by rises and falls in tide or lake level. It is that part of the river where the active extension of a floodplain is in progress by the deposition of the fine silts, muds and floating detritus carried.

Not every river has all three phases represented in its course whilst, on the other hand, a river which flows through one or more lakes may have all phases duplicated.

It will be possible to observe all these phenomena all over the project, if not in full size, in miniature. Every puddle and runnel on a road, every drain, every bit of erosion, will conform. Understanding them, and the mechanics of meandering, will help in controlling the undesirable aspects and in turning to advantage the beneficial aspects of the process during the construction and upkeep of the project's roads.

Meanders and course changes

The mechanics of stream flow as it affects meander formation and course movements are as follows:

- In a straight river bed, the fastest flow occurs at the point furthest from the bed of the river, where the frictional resistance of the river bed and resultant eddying slows the water down. Such erosion as occurs is therefore fairly evenly spread over the river bed (see Diagram 70 on page 214).
- When the water from a straight section of bed moves into a curved section, the fast flowing water in the centre persists in travelling in a straight line until it meets the outer bank of the curve, running, as it does so, over and across the the deeper slow moving water in the bottom of the river bed. As it meets the bank it is forced along the bank and downwards in a corkscrew motion and erodes the outer bank and loses some speed. It then moves diagonally across the bottom of the bed towards the inner bank, losing still more speed, and depositing the products of erosion opposite and downstream from the point from which they were removed (see Diagram 71 on page 214).

The cause of the one-sided cross-sectional shape of a river bed on a bend compared with the shape on a straight sector is thus explained. From this can be deduced the fact that bends both tend to enlarge their radii and to move downstream. This process continues, loop expands into loop, individual loops are short circuited and become isolated by further deposition and form "ox-bow" lakes or ponds. The ox-bows eventually silt up and disappear over a series of floods and become just a part of the floodplain alluvia. Diagram 72 on page 215 shows in three successive drawings four steps from course A through B and C to course D including the formation of ox-bow lakes.

DIAGRAM 70   Water flow in a straight river bed

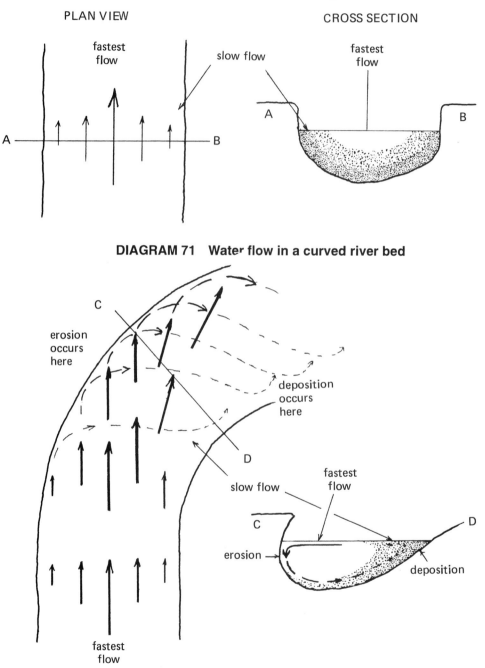

PLAN VIEW

fastest
flow

slow flow

A ———————————— B

CROSS SECTION

fastest
flow

A   B

DIAGRAM 71   Water flow in a curved river bed

C

erosion
occurs
here

deposition
occurs
here

D

slow flow

fastest
flow

C   D

erosion →

deposition

fastest
flow

DIAGRAM 72   The movement of meanders, course changes and the creation of ox-bow lakes

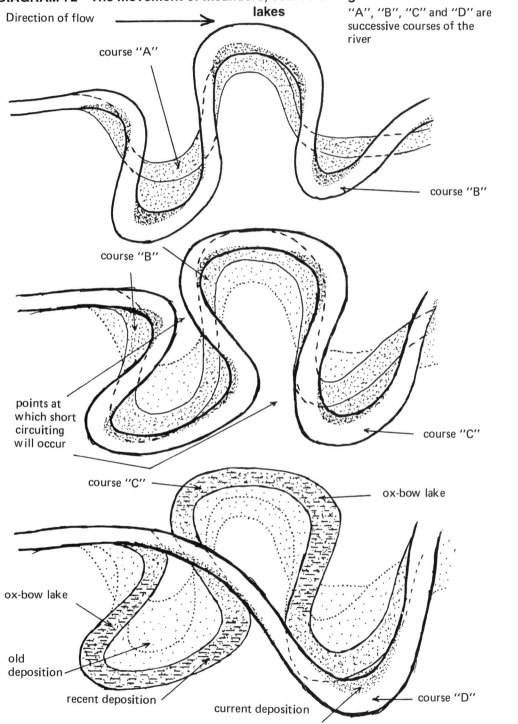

Direction of flow

"A", "B", "C" and "D" are successive courses of the river

course "A"

course "B"

course "B"

points at which short circuiting will occur

course "C"

course "C"

ox-bow lake

ox-bow lake

old deposition

recent deposition

current deposition

course "D"

A particularly active form of meandering will occur where a river carrying large quantities of boulders and gravels exits from a section of torrent course and first enters a floodplain. During flood times the impetus gained by the river is suddenly dissipated as it meets the resistance of material previously deposited or floodwaters already accumulated. The coarse materials in suspension are immediately dropped because of this substantial loss of flow speed, often right in the middle of the current bed of the river. Later, when the flood has subsided and normal flow is resumed, the mound of coarse material so deposited may well be too high for the water to surmount any longer. The water will therefore percolate through the deposit, or remain trapped behind it, until it is able to find a way around the obstacle, usually by cutting into finer material deposited earlier in the history of the floodplain or into the constraining foothills on either side. The effect is an alteration of course and a fairly rapid one at that. When the next periodic flood occurs the cycle will be repeated, the "new" course will be blocked and subsequently yet another found. In such situations the build up of a talus or fan of stones and rocks can be rapid, resulting in frequent course changes and the destruction of an everwidening area of any good soils in the way. Eventually a "braided" river will result. Such places are basically unstable and should be avoided when road making, if this is possible. They can however be a good source of roadstone – and the removal of the deposits for roadstone will ameliorate the meandering problem.

Methods of control

Avoidance

At the risk of being repetitious it is pertinent to emphasise the desirability of trying to avoid the need for river control. The careful selection of bridge sites and ensuring that roads are not routed near rivers that show evidence of past course changes will help. However, it is accepted that projects sited on alluvial areas have to live with the rivers that gave them the good soils upon which they are based. For them, avoidance is probably not possible but selection of the battleground may be. Study old and new air photographs if available and seek places that seem to have been stable in the recent past at least. Attempt to find crossing places where the alluvia are "pinched" between hills – the opportunity for major changes of course will thereby be restricted to some extent. Reduce the number of crossings to the minimum.

Straightening a section of river

It has to be remembered that in the process of straightening a river one will also shorten it. If the fall of the river was, for example, 1 in 500 (which gives quite a respectable speed of water flow), straightening a sector containing two or three near ox-bows could result in a reduction of length to one third or even one fifth, and corresponding falls of 1 in 166 to 1 in 100, in this example. Such markedly steeper gradients would induce far higher flow speeds and increase the erosion capability of the river. Doing this sort of work therefore is not without its risks and, unless weirs are built, it will usually be the case that when it has been done there will be long

DIAGRAM 73 Straightening out a section of river bed

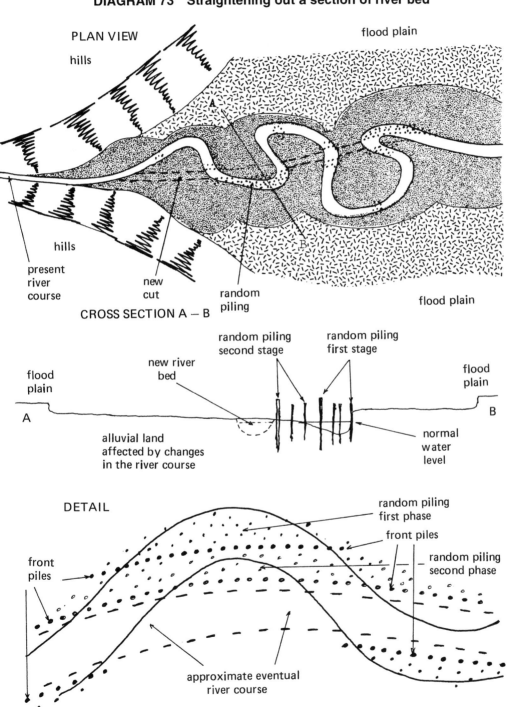

PLAN VIEW

flood plain

hills

hills

present
river
course

new
cut

random
piling

flood plain

CROSS SECTION A — B

random piling
second stage

random piling
first stage

flood
plain

new river
bed

flood
plain

A

B

alluvial land
affected by changes
in the river course

normal
water
level

DETAIL

random piling
first phase

front piles

front
piles

random piling
second phase

approximate eventual
river course

term effects, both downstream (greater deposition) and upstream (greater erosion), that will require monitoring. On a small scale it may be reasonable to put a bulldozer into a stream which has a firm gravelly bed and just cut a new and straight section of channel with little in the way of serious after-effects. On a larger scale much more care must be taken.

A method which takes time, but which will enable one to gradually reshape and control a river flowing through recent alluvia – without the risk of things going radically wrong – is to map the course of the river and of the land it has eroded away in its previous course changes (see Diagram 73 on page 217) and then place a series of cuts which will result in a straighter but not necessarily completely straight river. These can be effected, one by one, over a period of months by digging a trench, not necessarily as wide as the river, but more or less as deep, through which the water is able to flow. As each cut is effected, random piling (see below) should be put in at points along the old stream course that is to be replaced and below the cut where the redirection of the flow of the river is likely to cause unwanted erosion. The idea of the random piling is that it will slow down, but not stop the flow of the river through the old bed. Deposition will then occur in the old bed, while the river itself will widen the new cut. Some of the soil eroded from the new cut will be redeposited by the river amongst the random piling placed below the new cut where the erosion has to be controlled. When the process is seen to be working steadily and the old bed is seen to be silting up satisfactorily, the next cut can be made and the process repeated. In this way, cutting one bend at a time, working upstream, one is able to monitor the results over several seasons and do as little or as much as it is necessary and safe to do. One will not be landed overnight with a catastrophe.

Note that short lengths of revetting, sheet piling, gabions and other structures that simply divert the flow of the water away from a trouble spot are not recommended – anything that diverts the flow usually speeds it up, and wherever it is redirected to it can cause yet more damage. If these sort of things are used, then they have to be used extensively to provide a continuous and impregnable wall to be effective – and that is expensive for one is then contemplating the virtual canalisation of the river. Control of meandering should at all times be attained by slowing down the the river flow and allowing silt to accumulate and reeds and other vegetation to become established between the piles.

It may sound tedious but random piling, as a method of safely straightening a river, is cost-effective.

Random piling

The operation of random piling for river control demands intelligent and creative supervision but the materials required – jungle, eucalyptus or mangrove poles – are usually readily available or can be grown. In general, the piles should be long enough to be driven to refusal into the river bed and stand somewhat below the level of the surrounding ground. At flood times, therefore, they will be submerged and flotsam will pass harmlessly overhead but, at times of normal flow, they will stand well clear of the water. The spacing of the piles should be between six and ten pile diameters apart and, except for those piles edging on what is intended to become the new river fronting, truly random in their placing (see Diagram 73). Those piles forming the frontage should be heavier and stronger than those behind.

The wood used for the piling does not need to be especially durable, except where the piles are likely to remain exposed to the air as, for example, when forming a frontage. The piles do need to be straight, properly pointed four-square to prevent wandering, and cut off square at the top. Bindings of wire or metal rings may be required to prevent splitting under the hammer for the larger piles but fencepost sized piles can normally be driven with a maul or metal tube driver without such precautions.

When dealing with big rivers it should be noted that control of bank erosion by random piling is particularly effective where the river is wide and shallow and has numerous shifting sand banks. The deeper and faster flowing the river, the larger the piles need to be, and this may require the use of a substantial pile driver. The method is still effective but the operation moves rapidly into the realms of major civil engineering works and out of what one might properly regard as estate work.

QUARRYING

General

Before any quarries are opened up it is essential to check whether any special permissions or mineral licences are required from the government. Royalties may have to be paid on stone or gravel won but most governments will allow use without payment provided that the material is used in the development of the project itself. Once it is sold outside or taken off the project, royalties become due. This is fair and a reasonable position to negotiate for.

As quarrying is often regarded as a dangerous activity, there may also be special regulations designed to ensure the safety of personnel working in the quarry. Information on these, if they exist, will have to be sought and the relevant sections complied with.

Find out what, for the purposes of the government regulations, constitutes a "quarry" as distinct from a "borrow pit" or "earthmoving". Usually excavating a borrow pit or earthmoving in the process of road building will not be regarded as operating a quarry. Loading gravel recovered in the course of making a road cutting for use on another part of the road may also not be regarded as quarrying for the purposes of the law. Check first: this might save or reduce bureaucratic interference.

When effecting these enquiries, make sure that both central and local government bodies are contacted and that the final agreement is accepted by both.

Gravel quarries

River gravels and beach shingles

Beach shingles should only be used if they can be removed without destabilising the coast line. Once destabilisation occurs quite catastrophic damage can result to beaches and the facilities and installations on them. Live coral reefs should not be used to provide roadstone, their destruction too can destabilise a coastline.

River systems can be similarly destabilised but the results are usually more easily curable and under some circumstances, as noted earlier in this book, the removal of gravel banks from a river bed can be positively beneficial in saving fertile alluvia at the river margins from erosion and maintaining the river in an acceptable course. This is particularly true of braided rivers.

Both these activities may deplete, change or destroy local fish stocks and, in the case of rivers, will cloud the water, affecting anyone using it downstream. These matters have to be taken into consideration and perhaps arranged before work starts if unnecessary friction and unpleasantness with neighbours is to be avoided.

Do remember that equipment working in sea water will be at considerable risk of accelerated corrosion and should be thoroughly washed down with fresh water immediately upon leaving the sea. If working in a river that can flood heavily, make sure that all equipment is moved well clear of floodable ground after each day's work.

Work on beaches will be affected by the tides and may be impossible if the shingle becomes covered. Shingle beaches often have fresh water springs seeping through them and emerging well below the high tide level. These can sometimes act like quicksands under wheeled machines, so take care. Look for fresh water out-pourings at low tide. Make sure that equipment used has large tyres and can be promptly extracted if it becomes bogged.

If the river bed is really firm, and the water not too deep, it will be possible to scoop up gravel and load it in the flowing stream bed. However, proper articulated loaders and articulated dumpers are likely to be necessary to do this sort of work continuously and reliably. Equipment without four-wheel-drive and with small wheels and tyres will get itself stuck. Alternatively, a bulldozer can be used to scoop gravel to the water's edge and from there it can be loaded into tippers for delivery. If the water is too deep for a bulldozer a dragline may be required.

Should the material be very ill-graded for size, a simple screening plant can be used and the gravel separated into sizes that are immediately usable on the roads and large stones that will have to be crushed or be discarded.

Gravel beds

The recovery of gravels from gently sloping ground will necessitate no more than the stripping of the overburden cleanly before work can commence. If the site is so level that drainage may become a problem the establishment of drainage and satisfactory access within the quarry for

DIAGRAM 74   Ramp for loading gravel by bulldozer

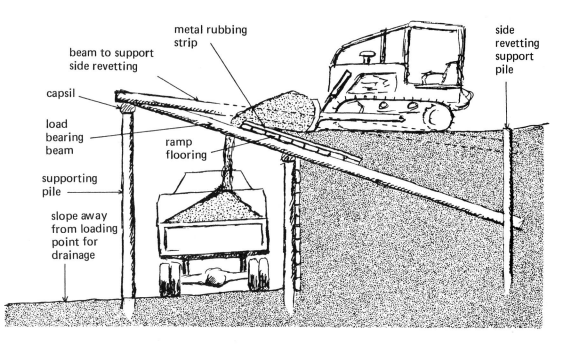

metal rubbing
strip

beam to support
side revetting

side
revetting
support
pile

capsil

load
bearing
beam

ramp
flooring

supporting
pile

slope away
from loading
point for
drainage

DIAGRAM 75   Operators must not be allowed to create overhangs

DIAGRAM 76   Crusher site showing two levels

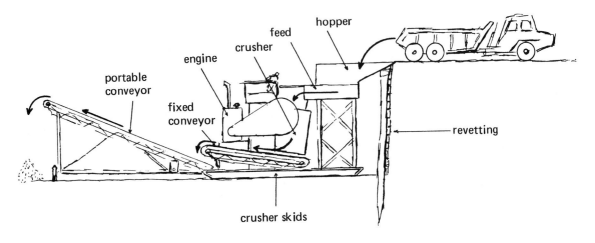

portable
conveyor

engine

fixed
conveyor

feed

crusher

hopper

revetting

crusher skids

DIAGRAM 77   Accurate surveys are required for drilling to get correct depth for the future quarry floor and to show where rock is thin and charges must be reduced

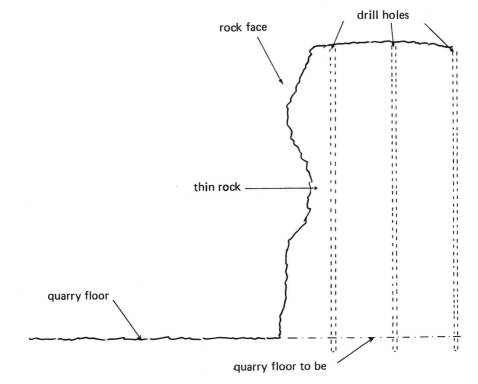

rock face

drill holes

thin rock

quarry floor

quarry floor to be

the dumpers may be necessary before any quarrying starts. If the gravels have to be quarried wet from the deposit a tracked backhoe may be a more appropriate tool for the job than a bulldozer and/or front end loader. The upkeep of effective drainage will then be as important as the maintenance of an access road for the dumpers.

Outcrops of gravels and binders

Whereas gravel beds will tend to be of unconsolidated recently deposited materials, outcrops are likely to be much older, geologically speaking, and harder, unless derived from recent volcanic activity. The boundaries of a (geologically) new volcanic ash fall are likely to be clearly visible. The ashes will have little or no overburden needing to be removed, will be loose and easily loaded by front end loader. Older deposits will have weathered and, unless the softer products of weathering have been removed by erosion, covered by an overburden. This needs to be sampled through before a quarry is established to find out the shape and size of the deposit. If the result of this sort of survey is satisfactory, then access should be provided and clearance of the overburden effected in such a way as to permit quarrying to commence at the top of the deposit, with loading a convenient distance below it. If a bulldozer with rippers is to cut the material initially, then the bulldozer's work will be facilitated if it can rip and push downhill and heap the gravel at the bottom for loading with a front end loader.

Loading can be done directly by the dozer off a ramp (see Diagram 74 on page 221 and Picture 9.1 on page 230) but this will mean having the dozer in the quarry every time loading is effected. This may not be convenient and is less flexible than having the bulldozer stockpile cut gravel for loading by front end loader as and when required. The drawback of this method is that the loose gravel will soak up rainwater to some extent. With gravels that are free or nearly free of clay binder this is of little matter. The dozer operator can "waterproof" a heap of raised coral, laterite or gravel containing binders by shaping and consolidating it a bit to shed the worst of any rainfall. For gravels containing a lot of binder and, of course, for clays being cut for use purely as a binder, this is quite unacceptable – these materials must be cut and used immediately.

Over a period of time the quarry will remove the hilltop or dig into the side of the hill containing the deposit. IT IS MOST IMPORTANT THAT IN THE PROCESS OVERHANGS ARE NOT CREATED (see Diagram 75 on page 221) which can collapse causing injury and damage. Unskilled operators with front end loaders and backhoes are likely to create overhangs and must be warned against this danger. Bulldozer operators may produce high vertical faces which can be nearly as dangerous and so they need watching too. The upper edge of the quarry should be sufficiently far back that a stable excavation slope, certainly not steeper than 1 in 1, shall result.

Rock quarries

General

The operation of a rock quarry will normally involve the use of drilling equipment, explosives, rock crushing and perhaps screening plant. Occasionally an outcrop of rock suitable for crushing will be found that is sufficiently fractured naturally, to be torn down with or without prior ripping, and loaded by a face shovel or backhoe with rock bucket. Even though this will obviate the need to use explosives, stone crushing is still a far more complex operation than simple gravel quarrying and it is assumed that any project going in for this seriously will obtain staff experienced and qualified to run such a quarry. However, so that project management shall have some idea of what is involved and what the quarry operator's objectives and problems are, a brief sketch of the principles is given below.

Opening the quarry

The objective is to produce roadstone of the qualities and at the quantities required from a given rock outcrop as cheaply, safely and with as little environmental disturbance as possible. Therefore operations involving the use of explosives will require:

- A thorough survey of the outcrop and its overburden to define how much material is likely to be available and how much overburden has to be removed to get at it. From the information obtained during the survey it will be decided how to get access to the deposit and how to work it. The objective will be to work in such a way as to provide a good level quarry floor for the machines loading and transporting the shattered rock to the crushing plant. Simultaneously, safe access to the top of the face for the drilling machinery which will prepare holes for the explosives must be retained. The drillers will operate by creating a series of benches as deep as the maximum drilling depth that it is intended to work to initially, probably with jackhammers, and later with mobile drilling rigs. The jack hammers can be used efficiently to a depth of 3.5 metres but beyond that depth, mounted drills become necessary. For holes up to 21 metres (70 feet) deep 50 to 76 millimetre drills are required and for holes of more than that depth, 100 millimetre drills will be required. Two or three jack-hammer benches opened when starting the quarry can be reduced to one mounted drill bench at a later stage.
- The stripping of the overburden, at least as far as is necessary at that time, to enable the face to be worked. The overburden can probably be utilised to provide an adequate catchment ledge for the rock to fall on and form the beginnings of the quarry floor.
- The construction of access roads at safe gradients between the benches and the face and between the face and the crusher site.
- The siting of the crushing plant as close to the face as possible, consistent with safety, and preferably downhill from it. Unless the crusher is to be fed by conveyor belt or some other means of lifting the stone to the jaws, the site should have two levels (see Diagram 76 on page 222): a high level at which the shattered rock is delivered to the crusher's feed tray by the dumpers carrying it from the face and a low level where the crushing plant will stand and the crushed stone will be stockpiled. The interface between these levels where the crusher will stand will require a strongly constructed revetment of timber, gabions, steel shuttering or such like, to hold it up. IT IS ESSENTIAL THAT AMPLE ROOM BE PROVIDED FOR MACHINES TO MANOEUVRE AND FOR

STOCKPILING TO TAKE PLACE AT BOTH LEVELS. Failure to provide sufficient room is a common fault and hinders efficient operation. Use the overburden if it is suitable, and if there is enough of it available, to help build up an adequate area at the upper level.

In addition:

- Effective drainage may be required: there is no excuse for machines wallowing around the sites up to their axles in water and mud or for the waste of valuable stone being used by operators to try to correct the situation.
- Effective fencing may be required to keep people out of the quarry for their own safety.
- If your supplier cannot deliver explosives on the day and at the time required and cannot remove any excess at the close of operations, the construction of explosives magazines may be required.

Operation of the quarry

Where the rock is naturally fractured and can be extracted and loaded by machine for delivery to the crusher, operation at the face will be similar to that of a normal gravel outcrop quarry. Operating conditions for the machines involved will, however, be much harsher. Where the use of explosives is required, the preparation of the face by surveying, drilling and blasting imposes upon management the need to plan ahead to ensure continuous production.

The business of drilling can take days, even weeks, depending on the equipment available, the extent of the pattern to be drilled and the constraints to blasting that may be imposed by the authorities. It will pay to get this work well in hand long before the crusher is installed, or if several quarries are being served by a mobile crusher, long before the crusher is due to arrive.

Besides the crusher and screening plant the quarry will require:

- A face shovel or backhoe with rock bucket for loading the raw material.
- Possibly a rockbreaker head for the backhoe to break up rock that is too large for the crusher.
- One or more dumpers with rock bodies to carry the coarse rock to the crusher.
- A front end loader on wheels to load either the crusher and/or the tippers that will deliver the crushed stone to the roads.
- A shed in which tools can be locked and in which records can be kept.

Explosives (see Pictures 9.13 to 9.18 on pages 230 to 238)

The choice of explosive and size of hole must be appropriate for the rock type to be shattered. The scale of operations may also affect the choice. Expert advice on the particular site being considered should be obtained from the explosives supplier. From this advice can be deduced the type of machinery required to do the drilling, the spacing of the holes, the amounts of explosive required over a period of time to sustain production at the required level and the frequency of blasting. This last item may be modified by the need to comply with police requirements for the control of explosives: compliance might be so troublesome and expensive as to make it desirable to have a few very large blasts rather than frequent small ones.

The shot firer will decide where the holes will be and how deep they will be drilled. For this he will need accurate surveys prior to each blast which will enable him to calculate the depths of drill holes required to create an even quarry floor. With knowledge of the shape of the face he can provide the right amount of explosive correctly positioned in each of the edge holes to prevent stone either being sent flying dangerously far or not being adequately shattered (see Diagram 77 on page 222). His aim will be to shatter all the rock to a size acceptable in the crusher's jaws and yet do little more than cause the face to collapse down upon itself into a heap at the bottom, ready to be gathered up by a loader.

The following notes are provided to enable management to understand to some extent the terms used by, and the requirements of, the shot firer:

- In drilling measurements (see Diagram 78, opposite):

    - the thickness of rock between rows of holes is called the "burden",
    - the distance between holes in a row is called the "spacing",
    - the depth of the face from the point at which the hole is drilled to the floor of the quarry is called the "depth",
    - drilling to below the depth of the quarry floor is called "toeing" and is done to ensure that a relatively smooth and level floor can be made as the loose rock is cleared, and
    - the arrangement of the drill holes is called the "pattern".

- Burden is scaled to hole diameter, usually 20 to 40 times the diameter of the holes. Spacing distance is normally similar to burden distance. Stemming (the material used to seal in the explosive when the hole has been charged) depth should be 0.8 to 1 times burden distance and toeing (also sometimes called "subgrade") depth should be 0.2 to 0.3 times burden. From these measurements can be calculated the volume available for the explosives to be used.

- Theoretically speaking numerous small holes should give a better shattering effect than few large ones but:

    - holes of less than 76 millimetres (3 inches) diameter will not be able to take the newer and much safer slurry and emulsion explosives which come mainly in 70, 85 and 100 millimetre ($2^7/8$, $3^3/8$ and 4 inch) diameter sausage-like packs; this will necessitate reversion to the use of gelignite with or without "anfo" (prilled ammonium nitrate mixed with fuel oil),
    - the smaller the holes the lower the efficiency of the explosions,
    - drilling time will be markedly increased, and
    - the smaller the diameter of the hole to be drilled the less accurately it can be placed.

DIAGRAM 78 Drilling for blasting

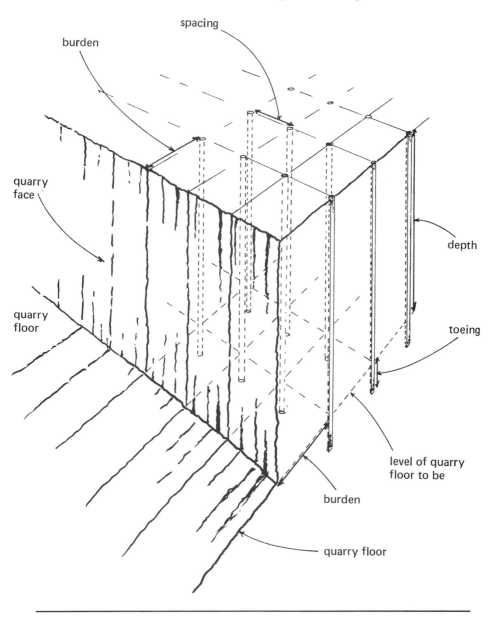

- A staggered (triangular) pattern of holes will give better shatter in massive rock than a square grid pattern. Where the rock needs only to be shaken down rather than shattered, burden thickness can exceed spacing distance.
- The drill dust from a wagon drill should be collected and not allowed to blow around. It can make life very unpleasant for the operator and does the machine no good. The drill

DIAGRAM 79   Sequential firing pattern to get effect of triangular spacing on grid drilling pattern

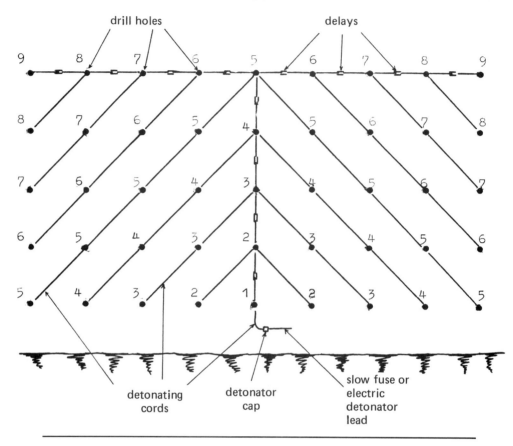

dust can provide chips for stemming in the holes later when they have been charged with explosives.

● After drilling, and before charging the holes with explosives, a wooden dolly of the diameter of an individual charge should be run up and down the holes to see that there is no obstruction to the correct placing of the charges (a rotan rod is ideal for attaching the dolly to). If a hole is blocked and it cannot be cleared with a compressed air line, it is better to redrill alongside it than to fire it with an incomplete charge. This is in order to avoid a lump of unbroken rock projecting from the quarry floor where the charge did not blow up the toe. Such a projection, once uncovered, would have to be separately drilled and blasted off in order not to inconvenience loading operations.

● Unlike nitro-glycerine based explosives, modern slurry and emulsion explosives are insensitive to friction and impact, are water resistant and do not cause headaches. They tend to heave rock rather than send it flying and to achieve this with a more even shatter. They require a minimum hole size of 76 millimetres (3 inches) and therefore can only be used where mounted drilling equipment is available. For small diameter holes the use of more sensitive explosives like gelignite or dynamite is necessary – this is because the size of the hole is very important to the efficiency of the explosion. The larger the diameter of the charge, the more effective it is, and the safer slurry and emulsion explosives do not reach their optimum yields at the smaller sizes. Nitro-glycerine based explosives are more unstable and therefore more dangerous to handle and to store. Nitro-glycerine can cause violent headaches, both by inhalation of the fumes and by penetration of the skin – it is very unpleasant to handle.

- Anfo (prilled ammonium nitrate mixed with fuel oil at the proportions of 94.5% to 5.5% by weight exactly) is a very cheap explosive. It can be used to eke out the more expensive gelignite in the smaller holes though with anfo too, larger holes increase efficiency. It is necessary to ensure that the anfo completely fills the hole without voids and that confinement is effective. The gelignite acts as the primer, the explosion of which ensures the detonation of the anfo. For small quarry operations the anfo can be mixed on site manually using, for safety, a wooden spade in an aluminium or wooden container. The best way is to use one 50 kilogram bag of the ammonium nitrate and 2.9 kilograms of diesel oil at a time. Until mixed, the two constituents are not classed as explosives and this can be an advantage from the point of view of avoiding interference with the logistics of the operation occasioned by the application of explosives regulations. In very dry climates over-rapid filling of holes with anfo can result in static electricity affecting electrical detonators and leading to premature explosions, but on a normal day in the wet tropics, the sort of small hand-filled operation implied by the note above is quite safe. Indeed, care must be taken to ensure that the ammonium nitrate is kept dry. It follows that it cannot be used in wet holes unless protected by loading it into polythene tubing, sealed at the ends like sausages, prior to loading into the drill holes.

- The explosives in the holes can be detonated either by electrically operated primers or by the use of detonating cords. If stray static electric currents are a problem it is safer to use detonating cords which are initiated either with slow fuses and detonator caps or electrically. Detonating cords are very noisy unless blanketed by a considerable depth of soil or turf (not less than 30 centimetres deep). This can be an objection to their use. They come in low, standard and high charge weights. The low charge weight is used for making the connections between holes to reduce noise, the standard is effective at detonating gelignite and some slurry explosives, whilst the high charge weight cord is required for some of the emulsion explosives.

- If primers are used they are lowered into the holes first and the charges dropped in after them. If detonating cords are used, the first charge is tied to the cord and lowered down to the bottom of the hole and subsequent charges dropped in after it. As the charges are loaded they have to be pushed firmly into place with the wooden dolly. When the last charge has been lowered, stemming is poured in and also tamped into place to completely fill the hole to surface level. Very often it will be found that the coarser fractions of the drill dust are excellent for this purpose. The drillings can be fed directly from the drilling rig into plastic bags to keep them dry.

Picture Group 9:   Quarrying

9.1   Loading dacitic ashes for road surfacing over a ramp of the type described in Diagram 74. For loading gravel from an outcrop when there is no front end loader available this is a simple answer to the problem. Standing on the loading framework is not recommended!

9.2   Jack hammer drilling to open a new quarry and establish a face. Scaffolding was necessary to get a good foothold on the slippery surface of the rock.

9.3   The use of jack hammers and scaffolding to open the first face of a new quarry.

9.4   The same quarry after two blasts. Note the quantity of large pieces of rock rejected as a result of using small diameter holes and gelignite. This had to be broken down with a rock breaker later.

9.5   A small mobile drilling rig, sometimes called a "wagon" drill, complete with air compressor which it tows with it, capable of drilling 102 millimetre (4 inch) holes.

9.6   Lines of holes drilled and capped temporarily with rocks marked with orange tapes to make their position easily visible. Note the bags of drilling chippings collected for use as stemming.

9.7   Tying detonating cord to the first charge to be loaded into the hole.

9.8   The first charge poised for lowering after the hole has been checked by running the wooden dolly down it on the rotan rod seen in the left of the picture.

9.9 Lowering the first charge on the detonating cord, note the use of two lines, this is a safety measure considered unnecessary by many shot firers. After lowering the charge is firmed into place with the dolly.

9.10 The second and subsequent charges are dropped down the hole on top of the first charge, each is gently firmed into place to expand the charge tight against the walls of the hole and the detonating cord.

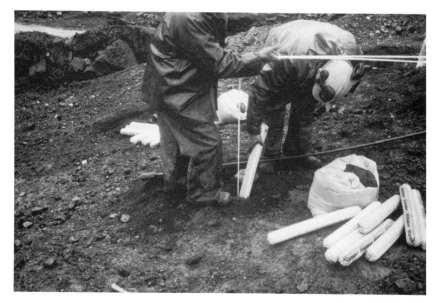

9.11   Firming down the last charge with the dolly.

9.12   The stemmings are tipped into the hole and tamped with the dolly, the last filling being firmed down with the feet – the boots are of soft rubber – and the tamping is completed. Note that the cord has been cut and is secured in place temporarily by the rock that was previously used to cap the hole.

9.13 Tying the charge lines to the main line of detonating cord. Note that the detonation will travel from right to left as shown in this picture, hence the direction in which the tie is made so that the detonation travelling along the main line will induce a detonation in the same direction down the branch line leading to the charge in the hole.

9.14 Another picture showing both the tying of the lines and a further three holes that have already been completed on the same line. Again, the detonation will travel in from the right of the picture.

9.15   Tying the electric detonator to the detonating cord. This is the point at which the explosion will start.

9.16   Checking the circuit before initiating the explosion. After this the warning calls are given and the shots are fired. In this case all charges were exploded simultaneously, no delays between lines serving the rows fired being used.

9.17 The results: the two lines of holes ready for firing; the white detonating cord connecting them shows up clearly.

9.18 This picture shows the effect of the blast. Note how little the rock has moved and the excellent degree of shatter, very few rocks will need to be broken down further before they are fed into the crusher. The pink painted sighting rod marks a survey reference point.

9.19./9.20 Breaking up rocks for the crusher the old way (picture taken in 1955). The skill of this old quarryman with a hammer on a whippy rotan shaft in breaking up large boulders of diorite after a few well placed blows was remarkable. Splitting this boulder in half took just five sharp taps.

9.21/9.22   Breaking rocks the modern way: 9.21 shows an excavator fitted with a hydraulic rock breaker tearing stone out of the side of a cutting during road construction. The product of this work was fed to the project's crusher subsequently. 9.22 shows a rock breaker mounted above the crusher's jaws to deal with anything too big that became jammed. It would have been better to mount it on the retaining wall than on the chassis of the crushing unit where it tended to wobble about in time with the motion of the crusher jaws. Basically a good idea.

- The use of detonating cords alone will result in all holes being detonated simultaneously. With large patterns there can be considerable advantages to putting millisecond delays into the system to get the sequential firing of holes. This will prevent the rock being thrown too far and can enable the rock to be thrown in a particular direction. A typical sequential pattern is shown in Diagram 79 on page 228. Other patterns can also be devised. Sequential firing will also reduce ground vibrations considerably.

- When all the holes have been charged, stemmed and connected up, the preparations are complete bar the attachment of either the capped fuse or the electrical detonator. This should not be done until the site is clear of all personnel other than the shot firer and his assistant. By this time all equipment that can be damaged should be off the quarry site or be shielded, and all entrances barred and guarded. The shot firer will then call the warning, using a klaxon or other audible device, and initiate the explosion. He must examine the results to check for, and deal with, any misfires before anyone else is allowed back to the face to work. When all is safe, he alone is responsible for giving the all-clear, whether immediately after the shot firing or at some time later when all misfires have been made safe.

Safety

The suppliers of explosives can provide very useful and complete information on safety. This should be thoroughly studied by management along with the local government regulations on the use of explosives. Management should then draw up a comprehensive set of "in-house" regulations and insist that they are adhered to. However, a few points are made here because they are the sort of dangerous mistakes that can be made inadvertently and which can have serious results:

- Always hold detonators by their necks, never by their heads.
- If electric detonators are used, do not use radios or mobile telephones to transmit calls from anywhere near the quarry during blasting operations.
- If nitro-glycerine based explosive gets spilt, do not scuff it underfoot. Friction causes it to explode very readily. Do not scrape it up. If it is sticky, use flour to sop it up with. Never slide or drop boxes of gelignite. Open boxes separately, well away from other explosives.
- Explosives and detonators must be stored in separate places. Store detonating cords with the explosives, not with the detonators. Detonators should not be transported with explosives or carried by someone carrying explosives. Do not carry matches or lighters with either.
- Do not allow ends of fuses or detonating cords to become wet, this can result in misfires and the dangers associated in dealing with them; seal cut ends with a dab of vaseline.
- Use only wooden tools to tamp with, ALL metal tools are unsafe.
- Keep hold of the detonating cords or detonator wires whilst the charges are being loaded and until the stemming has been tamped in place.

- Make sure that people know and understand the warnings that will be given at the time of shot-firing. It may be necessary to detail the signals on notices around the quarry to ensure that no one is left in ignorance.

This is not by any means a complete list of safety dos and don'ts.

After blasting

When loading is in process there will be some pieces of rock that are too big for use in the crusher. The operator of the loading machine should put these pieces on one side. Probably the cheapest way of breaking these up is manually, using sledge hammers on whippy (rotan) shafts. Another way is to blast them with plaster gelatine. This explosive is available in slabs for the purpose and is placed on the boulder on a patch that has been cleaned of dirt or moss. The quantity required is about 370 grams per metre (4 ounces per foot) thickness of stone. The charge should be completely covered with a wodge of thick clay before it is detonated. Alternatively, the hole can be drilled to a depth of about half the thickness of the rock and 50 to 100 grams (2 to 4 ounces) of gelatine tamped in and detonated. This method saves explosive but necessitates the use of drills. Both methods can result in shards of stone being sent flying for great distances and may be unduly expensive if security regulations make the use of explosives for this sort of work inconvenient.

ROAD SURFACING

Materials and methods

The objective of surfacing a road is to give it a surface which is more stable, harder wearing and less slippery than the material of which the main formation is made. If the road is being built in a material which has these qualities adequately, there is no need to apply an additional surfacing layer – make the surface of what is there.

There are two principal types of material with which a road can be surfaced, those with binders and those without. Crushed stone, shingles and oil palm shells have no inherent binding quality. Quarry gravels and rotten rock will usually contain some proportion of binder in them, and clay is used purely as a binder. Each material has its own particular properties in use:

- Crushed stone; this is hard, will provide a non-slip surface but lacks any binder. It should be crushed to less than 6 centimetres ($2^1/_2$ inches) in size and preferably the fines should be included in it, just as it comes from the crusher. If it can be supplied crushed to $2^1/_2$ centimetres (1 inch) at a reasonable cost, for most purposes this is better – it can be spread more neatly by the grader. If applied directly to a clay formation, this is best done when the clay is still wet after rain. It can then be spread and rolled in immediately and this will ensure that it is well bound by the clay. Well done, this type of surface can be almost as good as a tarmacadamed road and will last well for as long as the moisture content of the formation is correct. If the clay is hard, or if the formation is of a more silty or sandy texture, the crushed stone should be scarified in (i.e. the grader should scarify the road surface, mix the crushed stone into it by windrowing and then respread evenly over the road surface) and then rolled. It may be desirable to spray a little water onto the road surface in the final stages of the spreading, leaving it a while to soak in before rolling.

- River or beach shingle; like crushed stone, this will be lacking in any binder. It will have the advantage of being rounded which will make it much kinder on the tyres of the vehicles using the road but will also provide a marginally poorer grip in wet weather. Spread as for crushed stone.

- Oil palm shell; this is very light in weight compared with stone but comes in a size that is very suitable for road surfacing. It is very kind on vehicle tyres and surprisingly durable. Treat as for crushed stone.

CAUTION: materials that contain no binder should never be laid thickly on a road without either being admixed with the natural clay in the road by scarifying and windrowing, or being applied with clay as a mixture (two parts of stone, shingle or shells to one part of clay). Unstabilised, loose material, even if rolled hard, will be dangerous for motorcycles in particular and may make the steering of other types of vehicle difficult to control and so cause accidents.

- Quarry gravel; some quarry gravels will be derived from ancient beach or riverine deposits, others may be derived from igneous rocks that have been shattered (brecciated gabbro, for example) but remain in situ. The former will contain rounded stones, the latter angular material. Both are likely to contain some binder and fines. Another form of gravel, peculiar to the tropics, is laterite. It is formed by the chemical weathering and leaching of igneous rocks and contains large quantities of iron, and frequently manganese, as oxides, often in the form of pea-sized nodules. All these materials will probably be moist when delivered to the road for spreading and this is an advantage. They can be spread directly as a blanket of material over the prepared formation and consolidated immediately with a roller. The laterite will "set" after laying and provide a very durable surface. The use of a sheep's foot roller to provide a pockmarked surface on a newly finished formation, prior to the spreading of the gravel, will help to key it in place and make the grader's work easier: if this is not done, the freshly delivered gravel, which will probably be quite cohesive, can slide along a smooth rolled dry clay surface in front of the blade, making it very difficult to spread evenly.

- Rotten rock; consisting of partly weathered igneous or metamorphic rocks, or sedimentaries that have not been compressed to the stage of being really hard, it is likely to contain quite a lot of coarse material (see Pictures 3.13 & 3.14 on page 77) as well as binders. It will need to be spread initially by a light bulldozer which will be able to crush many of the lumps under its tracks. Further crushing with a compactor will be necessary and it may pay to use the road in its rough spread state for a while before grading it smooth. When it is graded, the grader should rip the road and the compactor recrush the material again so that the grader can get a satisfactory result. Rotten rock is very useful when surfacing the poorer soils as it gives much improved bearing capability to the road surface. Some rotten rocks can eventually become clayey on the surface and a little slippery during rain. It may then be necessary to give a thin surface dressing of crushed stone or fine river gravel. Raised coral beds (coral that has been raised from the sea in relatively recent tectonic movements) come into the rotten rock category if they are soft enough to be ripped and pushed out for loading by a bulldozer. Coral may not need the use of a bulldozer to spread it if the bulldozer operator in the quarry takes a little care to rip and cut it in such a way as to break up most of the lumps. A grader can then spread it with the aid of a compactor. The coral may need thoroughly wetting when being rolled after spreading in order to pack it down well and it may then set almost like a cement – do NOT mix it in with a clay soil, lay it on top as a blanket not less than about 5 to 7 centimetres (2 to 3 inches) thick and well compacted for the best results (see Pictures 3.11 & 3.12 on page 76).

- Clay; choose a red to chocolate brown clay with a good crumb structure to combine with crushed stone, river gravel, beach shingle or oil palm shells for spreading on formations made from the poorer soils. Mix at the rate of about one lorry load of clay to two of stone, windrow with the grader to get a good mix. This will give far better results on a non-clay base than applying the stone on its own, even if scarified in. Roads along beaches of coarse shingle can be effectively consolidated by dressing them with clay, much as one would dress a clay road with stone: scarify, windrow and spread.

It is not feasible in practice to spread these sort of surfacings to an exact width and thickness on top of the formation – spread them as a lens of material and accept that the layer will be thin on the edges, where few vehicles run anyway, and allow it to be thicker in the central strip over which most vehicles will pass. The practice of marking out and spreading to exact widths is only practical for sealed roads.

Attaining a perfectly smooth surface which is cambered or super-elevated to shed rainwater effectively and completely is important. Any small hole will hold water and, when the wheel of a heavily loaded vehicle passes over it, the water will be forced under very great pressure into the road formation. This is how potholes begin to form – the results for the road are as bad as a broken tile on the roof is for a house.

Organisation

For a given stretch of road it is easy to calculate the quantity of gravel to be put down by multiplying the length by the width by the depth and allowing for the volume of finally compacted gravel as being only two thirds of the volume loose, as loaded in the tippers. It is also easy to calculate the distance between tipper loads and to mark out with sticks where the loads should be delivered. If the grader is fitted with a front mounted blade there is no advantage in the tipper spreading the load as he drops it, the grader can cope with heaped loads. To facilitate movement along the road the heaps will usually be dumped along one side of it: if the road formation has a high camber, make sure that the tippers back across the road sufficiently that they do not become unstable when they raise their bodies to discharge the gravel – the writer has seen many tippers fall on their sides because the drivers forgot to take this simple precaution.

Check the time it takes for a tipper to load, deliver, discharge and return to the quarry and try to arrange that there are sufficient tippers working to prevent the loader from being idle between loads and the grader having to wait for gravel. The grader can probably spread far more quickly than the gravel can be delivered, so it is often better to dump the majority of the gravel on site before the grader comes to start its work. Dumping can be done at night; grading, if you want good quality work, cannot. On the other hand, rain, whilst not affecting shingle or crushed stone adversely, can make some gravels, some rotten rocks and all clays too wet to work, so these should not be dumped ahead if rain threatens.

It is likely that the grader can work faster than a steel rolled compactor. The compactor can work at night when the grader cannot so it can catch up that way or, if this is unacceptable, it may be necessary to provide two machines to get the best results. Alternatively, use a rubber roller which should have no difficulty in keeping up with the grader.

Have one or two labourers on site to dispose of the odd piece of rubbish or rock when the grader is working and to clear any outlet or roadside drains of any gravel or stones that fall into them during the operation.

Chapter 5

Maintenance

THE CAUSES OF WEAR AND TEAR

Rainfall and vehicle movements are the prime causes of wear and tear on roads in the tropics. The former will erode and soften, the latter will abrade and, where tyre loadings are too high for the condition of the earth of which the road is constituted, depress, deform and knead the road until it becomes plastic. Once in this state, breakdown is well advanced and it will take only a small amount of rain to destroy the road. During construction, considerable attention should have been paid to minimising the destructive effects of the rain by ensuring effective cambering, super-elevation and drainage, and to compacting the road formation well to prevent vehicles indenting the road surface. Care in the mixing in of gravels to ensure a dense smooth surface will have done much to reduce the effects of vehicle tyres scuffing the surface. But abrasion is inevitable, and the loose particles torn out of the road surface will be thrown away from the tracks of the vehicles. The immediate effect is to destroy the perfection of the camber or super-elevation (see Diagram 80a, 80b and 80c). As a result rainwater will become trapped and potholes will appear at the points indicated by the arrows if remedial action is not taken promptly.

The overloading of vehicles can contribute significantly to the destruction of roads (see Pictures 10.12 to 10.14 on pages 254 & 255). Prevention of this is an essential part of the road discipline that should be exercised to reduce the need for repair work.

In a situation where gravel is difficult to obtain and many of the roads on a project are ungravelled, it may be necessary to instruct drivers that all heavy vehicles should stop immediately when rain occurs and that journeys may only be resumed when the roads are dry again. This is not as Draconian as it may sound – in such circumstances the roads are likely to become far too slippery for lorries to drive in safety and so most drivers will understand the necessity. However, there will always be the odd one who considers that he needs to press on regardless until he can go no further. In the process he will inflict a great deal of damage on the road surface and may well get stuck or have an accident (see Picture 10.9 on page 253). Further damage may occur when the vehicle is extracted from wherever it has become bogged down. It is necessary therefore for the benefit of all concerned that the discipline is strictly enforced. To alleviate the discomfort of the drivers, light four-wheel-drive vehicles can be dispatched to collect them and take them back to base until the weather clears again. If rules like this are adhered to, a well formed road will dry out rapidly in the hot sunshine that often follows a rainstorm. Operations can then resume with the minimum of delay and with still good smooth surfaces to run on (see Picture 10.19 on page 258). The alternative is to experience the sort of breakdown of the road shown in Picture 10.16 on page 256 which was taken on a project where no such discipline was applied.

Not all road construction will be perfect and one of the most common failings is inadequate compaction. This is particularly likely to occur over culverts and should be watched for. As the road compacts with time and use, the filling over a culvert may become a hollow which, as vehicles run into it, is subjected to higher than usual pressures as the vehicles bottom out. Such a hollow can fill with water, which can only aggravate the problem, and can lead to distortion of the culvert. If signs of this sort of failure are seen, action to remedy the situation should be taken promptly to prevent damage to the culvert.

Blockages of drains, culverts and even bridges can cause considerable damage to roads. Frequently the causes of such blockages are trivial, a few leaves and twigs in a culvert, a small slide of soil into a side drain; sometimes they are substantial, trees and driftwood caught under

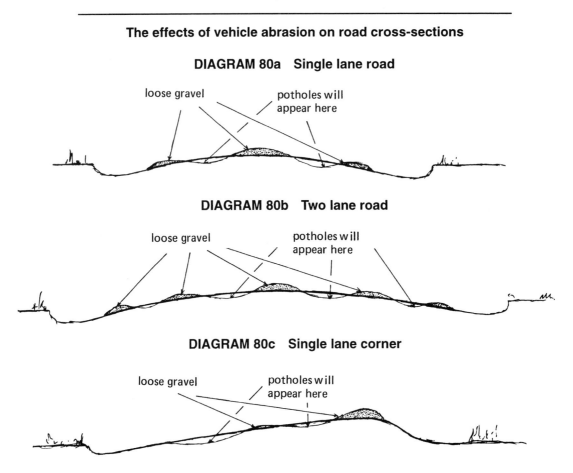

The effects of vehicle abrasion on road cross-sections

DIAGRAM 80a   Single lane road

DIAGRAM 80b   Two lane road

DIAGRAM 80c   Single lane corner

Picture Group 10:   Road maintenance

10.1   Not so much culprits as beneficiaries!

10.2   If roads are neglected, expect things to go from bad to worse very rapidly.

The importance of keeping bridges and culverts clear of obstructions is demonstrated in the following five pictures.

10.3/10.4 Although submerged, three culvert pipes are so badly blocked by wood that two pipes are hardly discharging at all whilst the third is only carrying about one third of its capacity.

10.5/10.6 At the height of the flood the water overtopped the stream bank and swept on past down to the next culvert, stripping off all the gravel from the side of the road as it did so.

10.7   It carried sufficient driftwood and rubbish with it to block the next culvert down.

10.8   The lining of drains with concrete is expensive and often ineffective. Once the water exceeds the capacity of the drain it will undermine the concrete and collapse will follow. This was remedied by filling with coarse stone, almost to road level, and allowing the water to percolate through the stone. The edge of the road was then dressed back to its original width with clay and crushed stone.

10.9 This road was formed and cambered during a break in between rains and was left to set. Before this could happen, a lorry driver broke through the barriers closing it off and tried to pass. Two others followed. It rained. The vehicles had to be extracted with the aid of a bulldozer. The resultant mess had to be hand-drained to empty out all the puddles and it then took several months to dry out before it could be reworked and the road made usable. Undamaged, the road would have set within a week or two, and could have been gravelled and compacted during the next brief dry spell.

10.10 Care needs to be taken not only in getting water to drain off a road but in seeing whether it may cause damage subsequently in the plantings the road is to serve. In this instance, discharge across unconsolidated alluvia newly cleared and planted to rubber resulted in an enormous gulley being created in a very short time during the rainy season. Considerable works were required to arrest the progress of this gulley and a lot of good land was lost because remedial action was not taken promptly. (Photograph courtesy of Peter Easton.)

10.11 The careful maintenance of embankments is most important. In this case the road has neither been graded out to the edge nor has a kerb with little outlet drains been made. The start of a little gulley can be seen in the right foreground and evidence of slumping following gulleying can be seen further back. To correct matters at this stage it would be necessary to lower the surface by a metre to get adequate width in firm soil to create a kerb or to grade off a wide enough shoulder to be safe.

10.12 This kind of rippling in clay can be due to either a lens of soft material under a road or by overloaded lorries. Immediate attention, before the next heavy rain, is necessary. In this case the road had been built across a swamp but had given no trouble until several overloaded lorries passed over it. Load limits were enforced and the road settled down once more.

10.13 The need to control axle loading as part of the road maintenance effort is further highlighted here. This road had been carrying 40 tonne loads without trouble for several weeks, when two lorries of the same type, came down overloaded to 60 tonnes. The rutting shown here was caused by this one passing of the two lorries. The picture was taken at dawn. The road was graded, compacted and back in service before the afternoon rains could further aggravate the situation. Soils: heavy tropical sandy clays.

10.14 This picture shows similar effects from use by overloaded lorries on clayless loess soils.

10.15/10.16   Even with no gravel on it a well-shaped and maintained road will dry out very quickly after rain provided that all heavy vehicles stop immediately the rain starts to fall. Failure to institute rigid discipline in this respect can lead to numerous accidents and a rapid destruction of the road surface. Once this has happened, further rain will be unable to drain off and the road will soon become a morass of impassable mud.

10.17 Maintenance of an ungravelled main road: sensible use of the rippers first makes reshaping of the camber very much quicker and easier. Note the texture of the freshly ripped surface.

10.18 Windrowing following ripping: note the fully loaded mouldboard blade, this makes for very high work outputs.

10.19   An ungravelled main road upon which strict discipline was enforced to prevent lorries moving during wet weather. As a result drying out was rapid, lorries could travel safely at speed, wear and tear on the lorries was reduced and output per lorry increased tenfold.

10.20   A well maintained oil palm estate harvesting road. This one is "gravelled" with palm kernel shells to give an all-weather surface. Fruit collection by tractor and two-wheel trailers.

10.21   A main road in a cocoa estate connecting harvesting roads 200 metres apart, one of which emerges in the left foreground. (Photograph courtesy of Anthony Herbert.)

10.22   A harvesting road on the same cocoa estate – pruners should not be allowed to leave their prunings in the roadside drains! (See also Pictures 3.1 & 3.2 on page 71 taken of a harvesting road on the same project, under construction 15 years earlier.)

10.23 The use of shade-tolerant *Desmodium ovalifolium* as an anti-erosion cover for roadsides, protecting a flight of hairpin bends.

10.24 A close-up of this very effective and easy to control legume.

10.25   A well maintained hairpin bend (not the one shown on page 79) that has been in use for 13 years. Note the bend has been kept level, the sharp drop away at 1 in 15 on the centre right of the picture. Also the clear, unsilted drain, centre left, capable of coping with any rainwater that could flow down from the road above. The ground is covered with *Desmodium ovalifolium*. Trees shade the road to maintain sufficient soil moisture to prevent it from becoming too dusty or suffering "dry breakdown".

This road links a cocoa estate with its head office and staff headquarters; it is used frequently each day, wet or dry season.

10.26   Periwinkle growing as a roadside cover. Although not quite as dense a cover as *Desmodium ovalifolium* it is frost resistant and easily controlled.

a bridge. The time to look for these things is either during or immediately after heavy rain when all the signs are fresh. Apart from obvious major damage or blockages, look also for runnels flowing along the road instead of across it and for small puddles that indicate where potholes will appear if action is not taken. Check embankments for erosion, particularly where they are in transition into cuttings; look for landslides both above and below the road; check small culverts, they are more likely to become blocked than the big ones. It is therefore recommended that THE PERSON RESPONSIBLE FOR ROADS SHOULD EXAMINE FIRST THE MAIN ROADS AND THEN THE HARVESTING ROADS IMMEDIATELY AFTER HEAVY RAINS AND ORGANISE REMEDIAL WORK PROMPTLY WHERE REQUIRED. Action taken speedily after one rainstorm can save major damage during the next, a heavy bill for repairs and possibly further costs resulting from the disruption of the normal flow of traffic.

ROAD UPKEEP

Maintenance maps

It will help considerably to make maps for maintenance purposes. They should show grades of road (main access road, main internal road, harvesting road), kilometre or milestones on all main roads, culverts, bridges, quarries and anything else relevant. Ideally they should be easy to photocopy without loss of clarity. They can be used by management and issued to foremen and lead operators. They will be invaluable in pinpointing troubles like new landslides; for showing a grader operator where he should go to work today, for example; and for supporting management's progress of works reports.

Maintenance of drainage

Bridges and culverts

All bridge and culvert structures should be the subject of periodic inspections for deterioration. Examine particularly the state of the joints of timber bridges – rotting of the wood nearly always starts there. Examine cement work for cracking that could result in weakening of the structure. Examine revetments, walls and foundations for voids and undermining by waterflow. Bare decked bridges suffer a great deal of hammering from vehicles at the ends just where the vehicles run off the gravel road and onto the wooden surface – check there especially for loose or cracked decking or runner planks. Look for sagging in culverts (see Diagram 39b on page 132). It is suggested that such examinations should be done annually as a special exercise, preferably in the early part of the dry season if there is one, to permit a programme of repairs to be worked out and implemented before the next wet season. Such inspections should be formalised with proper records.

In the short term it is essential that after every flood, bridges and large culverts are checked, cleared of any driftwood, and that any damage to piers or abutments discovered is made good

DIAGRAM 81   Feeding rainwater from a roadside drain into a small four-plank culvert on a side cut road

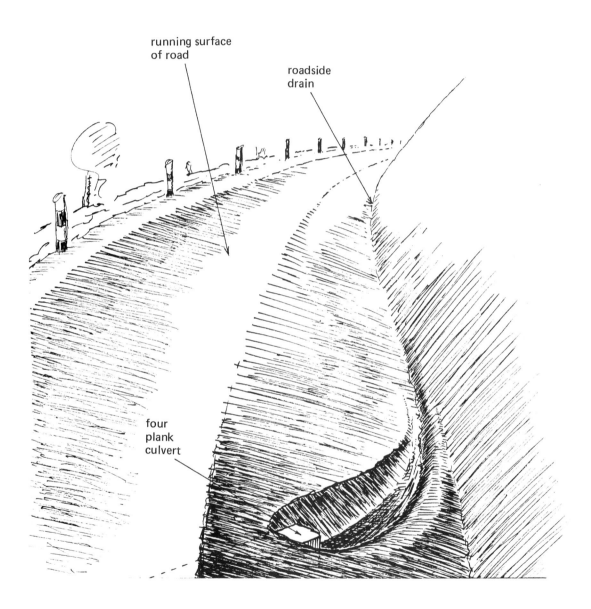

running surface
of road

roadside
drain

four
plank
culvert

promptly. Where the flood waters have been constricted by the structure, revetments and wings should be thoroughly searched to detect any signs that the water has created empty spaces behind them. Voids found should be filled immediately with suitable materials to prevent any further deterioration. Piled piers that have been well driven into the river bed are usually able to tolerate some deepening of the river bed without being seriously weakened, but look for conical pitting around the pile and, if found, fill the pits with heavy stone. Cast foundations that are not on solid rock, but have been dug into a stony or bouldery bed, can be undercut if the flood current has been fierce enough: a note should be made to examine them as soon as the water level has dropped enough for the purpose. If signs of undermining are found not only should repairs be effected promptly but consideration should be given to the cause. Perhaps there is need for additional flood relief capacity or for stream training to control or redirect the current.

Similarly, culvert grids should be cleared of accumulated driftwood and the upstream faces checked for damage. Look for any signs of water bypassing the culvert entrance, either by leaking around behind the wings of a pipe culvert, or running under the logs or foundations of a wooden culvert. If such signs are found then dig the affected place out and discover the full extent of the problem before deciding how to tackle it. The evidence may indicate that the culvert is of inadequate capacity and needs supplementing or replacing with a bigger one.

A void alongside a pipe culvert can be refilled with crushed stone or gravel well worked in, assuming that there has been no settlement of the pipes. If settlement has occurred the whole culvert may need to be lifted and relaid. Repair and reseal the concrete face and wings as necessary. If the culvert has no concrete face and wings, top off with good soil faced with wire mesh or old nylon fishing net and plant with grass or cover crops to reduce erosion. This will also help to clog up any small leaks and seal them naturally.

Downstream, the culvert outlet should be checked for the development of a hole underneath the discharge and for any tendency for the outlet to become undermined to the point at which it may collapse. Assuming collapse has not occurred, fill any hole with large boulders or rock as blasted ex-quarry until the end is properly supported. Create a bed of rocks capable of resisting the erosion of the discharge that will extend for a metre or two downstream from it. If some degree of collapse has occurred, dig out the culvert as far as it is affected and refill underneath with crushed stone or gravel, relaying the affected piping if necessary. In such a situation the construction of a concrete foundation and face wall under the end pipe may be necessary. Alternatively consider the use of a gabion for the same purpose. Cope with the outflow by placing coarse stone downstream, or the construction of a culvert-plate or concrete outflow channel to prevent a recurrence. But, it is the writer's experience that channels of cement usually crack and fail in time unless a well-founded reinforced structure is made. It is very much cheaper and just as effective to use coarse stone that the water can run through, dissipating its destructive power in the process, than to use either a smooth or a stepped channel. If needs be, use gabion wire mesh to help to hold the stone in place – being flexible it has a good chance of accommodating any subsequent settlement without becoming ineffective.

Wooden culverts, being usually built like small bridges, are less liable to be undermined and, in any case, their structural integrity will enable them to survive some degree of damage of this nature. If undermining is occurring consider whether there is a need for additional capacity.

This could be provided by installing another culvert alongside or by deepening the existing channel to increase its capacity, lining it if necessary with steel culverting or concrete. This can be an unpleasant task for the labourers doing the job but is perfectly practical unless the stream bed is very hard or rocky, or the culvert too shallow to permit access. Once the channel is lined the need to prevent a hole from developing under the discharge point will arise as with a pipe culvert. When making the lining, a skirt dug well into the ground at the inlet will be necessary to prevent the water from working its way under the lining. If capacity is adequate it may be that the undermining has been caused by an obstruction deflecting the water under the culvert beams or their foundations. Assuming little or no collapse it should be possible to remove the obstruction and refill the hole with coarse stone rammed in well. However, keep an eye on the repair for the next few months.

If a large proportion of the culvert has collapsed it may be necessary to make a temporary bridge. Use logs, reject lumber or coconut trunks wired together and earthed over, either on top of the existing culvert or to one side of it. If to one side, it is preferable to make it upstream of the culvert so that in the event of any ponding back from the temporary bridge occurring for any reason it does not affect the work in hand. For the smaller diameter culverts it is possible to use temporary steel bridging girders, one for each wheel track of the vehicles passing over the gap, and about three metres long. These can be made in the workshops and kept on hand for laying across the trench when a culvert is to be dug out.

Small culverts of any type are liable to be blocked by leaves and twigs and need checking frequently when wintering of the trees in the area sets in. Usually such small culverts only carry surface water, as for example under a side cut road, so that large volumes of water are rarely carried except for short periods during a heavy rainstorm. Where they are built into a side cut road, the discharge is likely to be onto fill and therefore a special eye should be kept on the outfall to ensure that any gullying resulting does not get out of hand. At the inlet end check that not only is the inlet clean but that the water is directed into the culvert by the drain (see Diagram 81 on page 263). Ensure that there is no chance for water to bypass one culvert and go on down to the next which would then, perhaps, be unable to take all the extra water. Such a situation, if it develops, can cause extensive damage to the road surface as a result of the water bypassing several culverts, gathering strength in the process, and cutting across the road at some point lower down. (See Pictures 10.3 to 10.7 on pages 250 to 252.)

Drains

The drains alongside main roads can to a large extent be maintained by grader. To do this really well an articulated grader (or one with rear wheel steering) equipped with a front mounted dozer blade, is desirable. The front dozer blade can be used to screef off grass and creeper growth and throw it well clear of the work before the cleaning and reshaping of the drain is done by the main blade. This initial screefing should be done with care as logs and rocks can be hidden by the weed growth. Where runoff drains have been constructed as shown in Diagrams 28 and 29 on pages 117 & 118, they too should be first screefed right to the very end unless, of course, crops have been planted so close to them that this is impractical.

By using the articulation to get the rear wheels right into the drain and the front wheels onto the road formation, the grader can get both a good purchase and good control for the grading of the drain to begin. The material from the drain should not necessarily be discarded, it may well pay to throw it onto the road – particularly if the road is due to be scarified – where it can be mixed into the existing gravel and replace some of the fines lost as dust raised by vehicles travelling over it during dry weather. When upkeeping runoff drains (see Diagrams 28 and 29) by grader, the turntable blade will not be able to finish the job as the front wheels of the grader will travel beyond the end of the formation. The operator should go as far as he can, then lift the turntable blade and back off, lower the front dozer blade and push the burden clear of the end of the drain. Done properly no further manual attention will be required. On harvesting roads the sides of runoff drains of this type may well have been planted over to the extent that the grader cannot do the job. In such a situation, once the grader has done what it can, manual labour will have to follow up and open out the runoffs immediately so that the water can get away during the next rainfall.

Roadside drains that cannot be cleaned by a grader should be checked prior to grading the road and any upkeep work or repairs required effected before the road is maintained. When the maintenance of the road by grader and roller has been completed the drains should be checked again and any material that has been dropped into them by the machines removed. Only then can the job be regarded as finished for that round.

On embankments that have kerbs as indicated in Diagrams 10a & 11a on pages 88 & 89 and pictured in Diagrams 21 and 22 on pages 100 & 103, attention to drainage will consist of cleaning out all the outlet drains by hand after the grader has finished doing the surface of the road. The outlets should have their floors cut just a centimetre or two deeper than the surface of the road at the road edge so that water can drop easily into them and will not be obstructed or diverted by fallen leaves, twigs or stones thrown up by vehicles. If this is not done, the outlet drains will get blocked very easily.

Maintenance of formation

Landslides

The basic formation and the land either side of the road needs to be checked regularly for deterioration. Look for cracks that could indicate the start of a landslide and for the beginnings of gulleying on embankments. Look for flows of water coming up against the footings of embankments and for springs and erosion appearing in the sides of cuttings.

Landslides can occur for a number of different reasons and have different forms and remedies :

- Conchoidal (shell-shaped) slide (see Diagram 82a on page 268); a type of landslide common on deep clays or recent alluvia. This type of slide can be greatly aggravated by the subterranean movement of water along an interface between an underlying impervious rock and an overlying pervious rock or soil. Spring lines will probably show on a face intersecting the interface between the two rock types and, where the road is formed in the upper material, should be regarded as a warning of possible trouble. Once started, conchoidal slides become self-generating in that rain water trapped on the ledge formed as a result will seep into the slide face, lubricating it so that it will slide more easily still in the future (Diagram 82b). Attempts to control this sort of slide must concentrate on draining off the water before it gets a chance to add to the problem and putting a counterweight (such as the gabion shown in Diagram 82c) at the foot of the slide to prevent it from progressing further. It is counter-productive to fill in on the top of the slide as this only adds impetus to the imbalance of the material situated on the slip plane. If perchance the road formation is on the slide, it is better to relocate the road well away from the edge. Make sure that no road drains, either from the old or new alignments, feed into the slide. If the road is at the foot of the slide and cannot be realigned away from it, it will be necessary to remove quite a bit of the material, possibly over a period of time, until near stability is reached. Then put heavy gabions, assisted by piles or revetting if need be, to retain the balance from spreading over the road (see Diagram 82c). Future problems can be ameliorated by the establishment of strong rooted covers to bind the slumped remnants of the landslide and to protect the upperface and the top ground that could be potentially the material for the next slide (see Diagram 82d).

- Bedding plane slide (see Diagram 8d on page 83); such slides are only likely to fall from above the road and usually it will be necessary to clear everything that has moved, if not all at one go, over a period. They are extremely difficult to hold and will continue to slide after each heavy rain or each earth tremor until all the material affected has fallen.

- Slide of unstable granular material (found typically in country showing sharp knife-edged ridges and evenly gradiented V-shaped valleys). Ill-consolidated ground lying at its natural angle of rest (like a heap of freshly tipped sand) will slide very readily from above the road once the road has been cut into it. Material thrown off the road will equally readily slide down far below it unless retained by vigorous vegetation growth. Such slides are likely to be very extensive when they start and the principal protection must be to retain a good thick cover of vegetation both above and below the road backed by retaining walls, revetting, gabions or netting as may be necessary. Such supplementary works are expensive and not always effective so there is adequate incentive for care to be taken of the covers. If such slides do occur, clearance with front end loader and tippers to remove the fallen soil to somewhere where it will not cause more trouble, is essential. Follow up with the re-establishment of covers as quickly as possible. If part of the formation has been lost it is almost always more effective to cut the road deeper into the hillside and take protective measures against further slides from above than to refill over the site of the slide. In an extreme situation it may be necessary and effective to bridge the slide if this can be done in a single span.

- Slide due to excavations below the road; this problem can arise on hairpin bends on steep hillsides where the road leading up to the bend can undermine the road continuing on upwards out of the bend. The writer has also seen such slides caused by the excavation of house or other sites too close below a hillside road. In such situations adequate support walling or revetting (see Picture 3.18 on page 79) should be put in at the time of

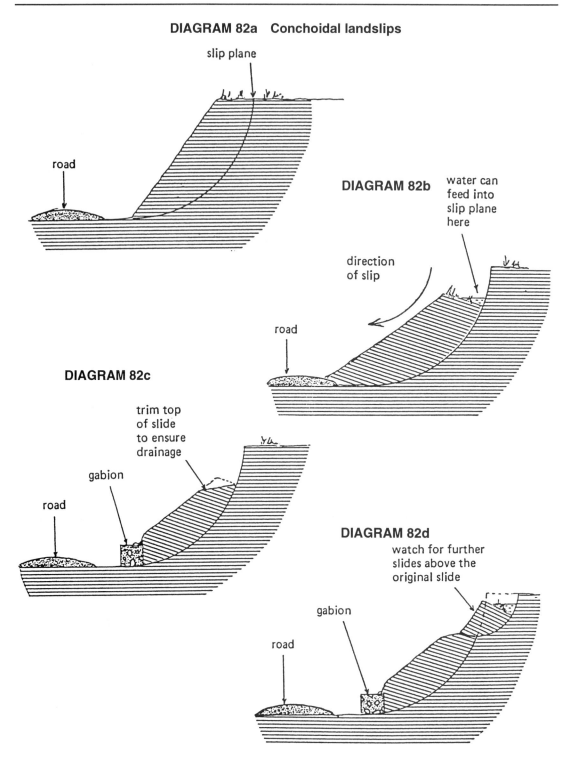

DIAGRAM 82a   Conchoidal landslips

slip plane

road

DIAGRAM 82b

water can
feed into
slip plane
here

direction
of slip

road

DIAGRAM 82c

trim top
of slide
to ensure
drainage

gabion

road

DIAGRAM 82d

watch for further
slides above the
original slide

gabion

road

construction. If this is not done and a slide occurs subsequently, repair will be expensive and inconvenient.

● Slides due to waterlogging; it can sometimes happen that small springs above the road (which have perhaps been allowed for during construction by the provision of a roadside cutoff drain and culverts), will cease to flow normally and will turn the ground through which they emerge into a slurry. This may creep down over a road, blocking all drainage in the process. The cause for this is not known to the writer but it may be a result of the felling of trees whose roots normally took up sufficient of the water to retain stability, or it may be a result of the vibration of heavy passing traffic. Whatever the cause, these sort of slides are among the most messy and difficult to deal with. Do not use a bulldozer to clear them, it is likely to get bogged down. A front end loader working at right angles to the road is more practical, at least to remove the bulk of the material from the road formation itself. A front end loader may not be able to reach in far enough to clear all the slurry that should be taken away to stabilise the situation. A backhoe, with its long reach and ability to re-excavate the drain, and perhaps enlarge it to provide some insurance for the future, may well be the only really effective tool for the job. Do clear enough material out of the way or the problem will return again and again until some sort of equilibrium is reached and the place stabilises. To help it stabilise re-establish covers as soon as possible.

Roadside vegetation

Control of vegetation and maintenance of visibility is an important aspect of formation upkeep, particularly on main roads. If labour is readily available, regular rounds of roadside slashing may be an effective means of control where machines cannot work and where visibility, for example at junctions or on the insides of corners, has to be maintained. Where labour is scarce or expensive, such parts of the formation that require that sort of care may have to be prepared so that vegetation control can be effected by a tractor mounted flail mower or chemical sprayer though, for environmental reasons, the latter method is becoming deservedly less popular. However, bear in mind the value of overhead shade in some circumstances both in reducing herbaceous growth, so enabling one to get a clear view underneath the tree canopy, and in preventing dry breakdown of the road in places where there is a pronounced dry season. It may pay to plant roadside trees or encourage the development of self sown tree seedlings.

Roadside erosion can be much reduced by the planting of suitable covers. The type of plant required should be able to grow well in both shady and sunny situations; neither grow too high nor spread aggressively over the road or up trees; provide a good dense shield for the ground from the effect of water droplets during heavy tropical rainstorms and have a good hold of the soil with its roots. *Desmodium ovalifolium* is one of the best covers for this purpose though it can be rather slow in starting (see Pictures 10.23 & 10.24 on page 260 and also front cover). The seed is minute, so that only small quantities by weight are required. It can be sown in mixture with other more commonly used covers and will eventually take over from them. If it is not available other species with similar characteristics should be sought, for example, dwarf periwinkle (*Vinca spp*) may be suitable where frost would kill off *Desmodium*.

Embankments

The upkeep of embankments made over soft ground will need special attention. Whether corduroyed or not, uneven settlement is likely to occur and this can result both in side slips of the edges of the embankment and unacceptable unevenness of the longitudinal profile. Provided that the roadside drains have been made at a reasonable distance from the base of the embankment (see Diagrams 11a and 11b on page 89), topping up the smaller side slips will provide no problems. If, however, inadequate distance has been provided the slip may well affect the drain too. In such a situation it is usually necessary to dig a new drain parallel to the damaged one but sufficiently far from it to be beyond the range of further slides. Only if this is not possible should piling and revetting be resorted to. It is expensive and, unless the vertical revetting sheets or piles are very well driven into firm ground, very liable to failure. The horizontal planking between piles must be taken down deeply enough or the fill will slip through beneath them.

Longitudinal unevenness can be countered by frequent grading, compacting and topping up with gravel until the position stabilises with time. This should not be neglected because the uneven settlement will mean that there are cracks and movement faces within the embankment and, if these get filled with rainwater, the condition of the embankment will worsen rapidly. The objective of the grading is primarily to maintain an effective water-shedding camber on the road. Prompt attention to outlet drains on the embankment is necessary where kerbs exist.

Cuttings

Look for signs of impending landslides and erosion and deal with them before the situation has become serious if possible. In particular, ensure that the steps of stepped embankments are functioning effectively as cut-off drains. Clear any deposits on the road edge that have fallen from above and threaten to block roadside drains before water gets ponded back behind them and starts to soften the formation.

Maintenance of the road surface

Grading and compacting

On straight roads both vehicles and rainwater will tend to throw the loose factions of the road surface downwards to the edges of the road. On corners, vehicles will tend to scuff the loose factions up to the outside of the corner and the rain will tend to wash the fines down to the lower inside of the corner. Although the two forces are in contention, on corners it will still be found that corners wear more rapidly than straights. If the corners have been adequately super-elevated the differences in wear rates should not be so much that the corners need special and separate early attention. If a corner shows a very high bank of loose gravel on the outer edge it is probably inadequately super-elevated for the traffic using it and the opportunity should be taken during a maintenance round to correct this.

Where the road is full of potholes, it should be ripped or scarified to the depth of the potholes before grading commences (see Pictures 2.7/10.17 & 10.18 on pages 47 & 257). If the ripped material has a good mix of gravel and fines there is no need to add more gravel. Windrow the surfacing from one edge to the other and back and respread, preferably with the roller working with the grader during the respreading.

If the road shows signs of "fatting" (this is where, on a clay soil base, gravel that has been used to dress it has sunk into the clay to the extent that the clay has come up through it to provide a thin surface mask that is slippery when wet) but is otherwise in good shape, a thin dressing of extra gravel or crushed stone should be added immediately after rain, spread by the grader and promptly rolled in with a smooth steel roller. If the road is in this state but ill-shaped, scarify to a depth of about 5 centimetres (2 inches) average, deliver extra gravel or crushed stone, windrow the whole from one side to the other and back to ensure proper mixing, spread and compact.

When starting to grade a section of road it is normal to do the roadside drains or road edges first, throwing the material towards the centre of the road. Even if the edges have been screefed beforehand and the majority of the vegetation thrown aside, it is still likely that some, particularly the very widespread and tough grass *Elusine indica*, will be present in the windrows of gravel produced. If a good quality job is required this must be thrown off by hand. The alternative, to throw it off with the grader, may result in the loss of a large proportion of the gravel. To leave it on the road is unacceptable.

It is not necessary to rip or scarify every time, indeed if the road has not been allowed to deteriorate to the stage at which potholes have formed, the majority of surface maintenance work can probably be done with the blade only. However, if it has become necessary to add more gravel to the surface, it is usually far better to scarify to enable the grader to achieve a good mix of the new and old materials and ensure that they bind well together.

The compactor should work with the grader as far as possible once the drains or edges have been cleaned and any scarifying and mixing done. When the grader starts to relay the surface it will be much assisted by having the roller compact, swathe by swathe, the work as it is done. The roller should direct its attention to the coarse material thrown off the grader's blade in the early stages of the operation in order to break up large lumps of soil and large stones. As the grader finishes off its work the roller should start to cover systematically the whole road width and continue this to completion. The grader should move onto the next section whilst this is in progress and do the screefing, drain and edge cleaning, and the scarifying before the roller arrives. These machines should work as a team supplemented, as may be required, by tippers supplying gravel or crushed stone to them.

During this process the grader operator should take care to effect the leading into and out of corners (see Diagram 18 on page 98) and junctions correctly and to maintain adequate super-elevation and camber on the roads. There is a lot of conceptual ability and operating skill required to do this work well and, in the writer's opinion, the really good grader operator who can be left with confidence to do the job well will probably be the most highly skilled, and should be the most highly paid operator of all.

DIAGRAM 83   Grading corners on gradients

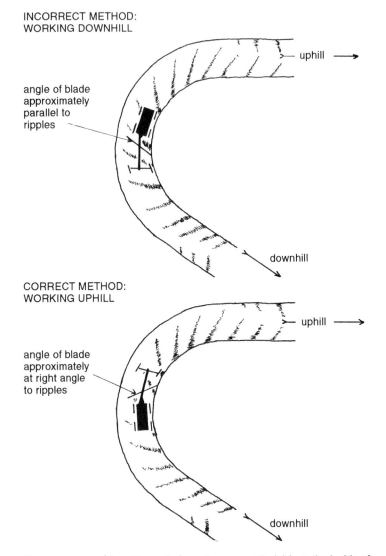

INCORRECT METHOD:
WORKING DOWNHILL

angle of blade
approximately
parallel to
ripples

uphill ⟶

downhill

CORRECT METHOD:
WORKING UPHILL

angle of blade
approximately
at right angle
to ripples

uphill ⟶

downhill

On a corner a grader will usually have to move material from the inside of the bend to the outside to maintain the super-elevation. Working down hill the blade, being parallel to the ripples will encounter alternatively hard ground and empty space to be filled. The grader will tend to bounce on its tyres in time with the ripples and do a poor job.

Working uphill the blade will be at right angles to the ripples and, cutting and filling several simultaneously, will do a smooth, even job.

DIAGRAM 84   Grading flights of hairpin bends

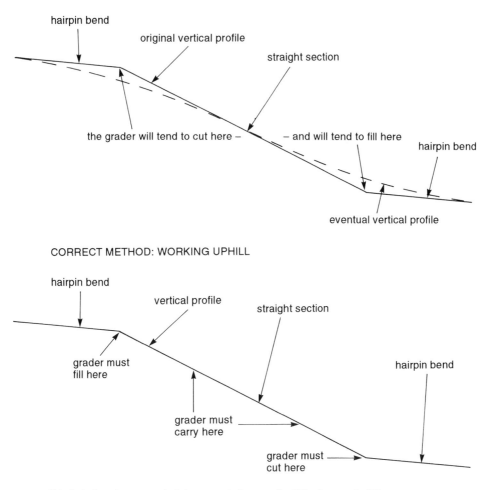

INCORRECT METHOD: WORKING DOWNHILL

hairpin bend

original vertical profile

straight section

the grader will tend to cut here –   – and will tend to fill here

hairpin bend

eventual vertical profile

CORRECT METHOD: WORKING UPHILL

hairpin bend

vertical profile

straight section

grader must
fill here

hairpin bend

grader must
carry here

grader must
cut here

If hairpin bends are graded downwards the result will be to round off the changes in gradient to something akin to the dotted line shown above. In practise, the rounding off of the hump will tend to precede the build-up of the hollow. This leads to the undercutting of the upper bend from below which will then rapidly become very steep and dangerous – particularly left hand bends for vehicles driving on the left, right hand bends for vehicles driving on the right.

Working uphill, the operator can make a conscious effort to cut and fill in the right places with the advantage of the material gained on the straight sector being on his blade to lay and buttress the bend above with.

Grading hills, corners and hairpin bends

It is acceptable to rip downhill with a grader but subsequent grading with the blade should always be uphill. This will counter the tendency of the gravel surface to be washed downhill by rain and so, in effect, one is putting the material back in place. However, this is not the only reason. On corners, rippling will tend to occur diagonally, not at right angles to the run of the road (see Diagram 83 on page 272). If the grader works uphill, the blade will be more nearly at right angles to the ripples whereas, if it works downhill the blade will be nearly parallel to the ripples. At right angles to the ripples it will cut them smoothly and do a good job; parallel with them it will be more affected by the alternate hard and soft ground (bear in mind that a grader is not on tracks which have no give, but is on tyres and so can bounce a little). This will not produce a good finish, with the result that the ripples will rapidly re-establish themselves in exactly the same positions they were in before.

It is even more important to grade flights of hairpin bends upwards. The vertical alignment of a set of hairpin bends consists of a series of inclines (between bends) connected by short nearly-level sections (the hairpin corners themselves) (see Diagram 84 on page 273, Pictures 10.23 & 10.25 on pages 260 & 261 and the picture on the cover of the book). It is important to maintain this vertical alignment. Working downhill, it is only too easy for a grader operator to cut too much immediately at the lower end of a bend and to fill immediately at the top of the bend below – the grader is only doing what it is designed for: cutting humps and filling hollows. If the operator works uphill he will more naturally split the operation into corners and straights, he will cut at the commencement of a straight and fill at the top end of it, so preserving the vertical alignment exactly as required.

Failure to preserve the step-like vertical alignment will lead very rapidly to the steepening of the hairpin bends, which will then become dangerous (see page 19). Worst affected will be vehicles travelling uphill, turning left if they drive on the left, or turning right if they drive on the right. This is because the helical form of a tight corner becomes more exaggerated the steeper it becomes. Thus, for example, a vehicle driving on the left up a left-hand corner can find all the weight of the vehicle is on the front-left and rear-right wheels and that, with the front-right and rear-left wheels virtually hanging in space, all drive is lost and the vehicle is unable to proceed. As a result, the vehicle will have to back down and try again, usually clinging to the outer edge of the corner regardless of the rule of the road, in order to be able to grip the road surface. The risk of an accident occurring is obvious: the fault lies with the person responsible for the poor road layout or upkeep.

Working uphill is not a problem if road gradients are reasonable. It only becomes difficult if gradients exceed 1 in 15.

Roadsigns

The upkeep of marker posts on corners, probably the most numerous and important of estate roadsigns, should not be neglected. They provide a guide for the grader on the insides of radiused bends and a warning as well as a guide for drivers on the outsides of bends and edges of steep drops. They should be checked routinely as road maintenance rounds are done and repairs effected. It is also advisable to make a special inspection periodically at which the need for further markers or signs can be evaluated in the light of experience over the past year. This can be a good rainy season job for then the ground is soft enough to hammer new posts in or to replace any that have been knocked over or broken. Painting of the black and white bands should be done twice a year in the tropics to ensure that the posts stand out clearly at night. Signs as such seem inevitably to become targets for stone throwing and shooting practice and it is often better to do without them unless the need is urgent. They will have to be very robust and also be repainted frequently. Metal letters, numbers and symbols screwed into stout waterproof plywood are the most practical solution as the repainting of the lettering to produce a legible result is then very easy.

Organisation of the work

In machine maintenance, the grader is the key unit in the team. The rate at which it can work and what it is capable of doing will dictate what other machinery and what labour will be required to work with it. The extent to which the road system is grader-maintainable is therefore very important, particularly if labour is scarce or expensive. It will have been noted from the sections above that there may be pre-grading work and post-grading work to be done in a maintenance round. This has to be organised properly so that machines and people complement one another's activities. It is desirable that the maintenance work should be the prime responsibility of one man, a Road Maintenance Team Foreman who should have the ability, status and transport to organise the day to day running of the operation. He will need one or more maintenance teams, dependent upon the size of the road system and upon the rate at which the roads deteriorate from the ravages of weather and traffic. The actual length of road to be cared for by a team will vary from project to project quite considerably depending not only on the constitution of the team, the traffic and weather, but also on the types of soils and gravels and the terrain of which and upon which the system is built. Unless all these factors are remarkably similar, attempts to base estimates of what is required or can be achieved on one project upon experience at another, can at best only be a rough guide. The same applies to comparisons of performance between one project and another: they can be thoroughly misleading.

Some notes on the forming of teams appear in Chapter 3 and on personnel in Chapter 7 and so are not covered here. However, it should perhaps be pointed out that the road maintenance gang should not be the haven for the project's "dead-beats"; the job requires intelligence and imagination both to conceive shapes and forms and construct them, and to be able to conceive what could go wrong if a shape is incorrect or if something does not get done when it should do. A feel for soil condition is essential and, amongst those with positions of authority within the teams, a feel for when jobs should or should not be tackled.

The ideal time to maintain the surface of a road is when the ground is moist enough to compact well but is not sticky or muddy. Work cannot proceed if the ground is too wet (the exception is the addition of crushed stone or shingle without prior scarification which can be done under quite wet conditions provided that the road base is sound). Work can proceed if the ground is dryer than ideal but should not do so if dry breakdown is going to result. Getting the conditions right is far more important on a busy road than on a little used road.

Grading can proceed on a busy main road with little traffic disruption if it is wide enough to allow traffic to pass easily, though there should be a hard and fast rule that ALL TRAFFIC GIVES WAY TO ROAD MACHINES THAT ARE WORKING – and that includes the project manager's car! (It is wise to ensure that all the road maintenance team machines are equipped with beacons and that they use them when working.) If conditions are perfect the traffic will assist greatly in the compaction of the surface. If conditions are too dry the fines will be turned to dust which will make the life of the operators very unpleasant and ensure that the road becomes terribly muddy when the next rainstorm comes; if too wet, the whole will become a morass of mud immediately. It is therefore very difficult to plan work on the basis of regular rounds at intervals of so many days or weeks. It is better to be flexible and take opportunities when they occur. Never allow maintenance to fall so far behind that, if a prolonged wet spell should occur unexpectedly, the roads could be in a real mess before the end of it. Frequent light maintenance of main roads is the best method.

The upkeep of harvesting roads can allow more flexibility especially if harvesting rounds are periodic. One can then arrange to clean up the roads (including drains) immediately the harvesting has been done. This is so that the roads shall be free draining, and can settle and harden off naturally, before the next harvesting round. Work can thus proceed when dryer than ideal in the knowledge that little or no traffic will pass that way before the next rain has fallen and the road has dried out again. This is important because an undisturbed wetting and drying like that will bind the surface very satisfactorily. Harvesting roads under tree crops will in any case tend to retain moisture better than exposed roads, so that they can be worked and compacted in dry spells when work on a main road would be impractical.

Forest crops, whether for timber or for pulpwood, are only harvested once every several years. Harvesting roads can be allowed to become grassed over between times, but cannot be entirely neglected. Occasional inspections of drainage systems, culverts and bridges are necessary because blockages left unremedied can result in serious damage to the road, either by washing out or by the softening of the formation. Prior to a harvesting round, herbicide should be applied to the road surface to ensure that the vegetation is not only dead but has substantially disintegrated, before a grader is used on it. This will make the operator's work easier and prevent gravel being screefed off and wasted. If the formation has been kept well-drained and in good condition, it is likely that preparation of the road prior to harvesting will consist mainly of compaction, the first round of which can be done, with advantage, prior to any work by the grader. Good drainage will ensure a hard formation, but inevitably grass roots will "fluff-up" the actual surface of the road. This needs correcting before loaded lorries get onto it and turn their trackmarks into ruts which will contain or channel water in the next rainfall.

During long dry spells some main road upkeep may be necessary to counter the effects of dust and surface rippling. If water tanks are available it will be beneficial to wet the road at nights to help bind the surface. Spraying should start at around sunset and, in the cool of the night, the water will soak in well before the next day's traffic appears. Wetting the night before a grading round is due will also help. If water spraying is not possible then grading to eliminate rippling should be very superficial and a steel roller should only be used with considerable caution to compact the ground afterwards. A rubber roller is better, it will compact without crushing and shattering the softer gravels to such an extent.

Long dry spells can lead to dry breakdown of the road, usually in patches. If these patches become unacceptably big and deep the only answer is to scoop out the loose material and refill with damp gravel or moist clay topped off with crushed stone. Like filling potholes by hand in the old-fashioned way, the patches should be filled proud of the road surface and tamped flat and hard. Shallow fillings will not last well but deep fillings, 20 centimetres (8 inches) or more, will hold for quite a long time.

If the project has a climate with a regular dry period of four months or more each year, then the provision of water tanks to spray the busier roads nightly is strongly recommended.

Long wet spells may result in the grader being unable to cope with the extra work required and potholes will form in the project's busiest roads. In such an emergency, a gang of labourers with wheelbarrows and a supply of crushed stone or shingle can usefully ameliorate the situation and reduce wear and tear on the project's vehicles but grading and rolling should be effected as soon afterwards as possible.

The maintenance of roads should be to the same high standards as the maintenance of the vehicles that run on them, they are, after all, complementary parts of the same transport system.

Priorities in emergencies

If the project is affected by a major climatic calamity such as a flood or hurricane, the primary priority must be to secure access to the road teams, their equipment and their fuel supplies. When that has been done, access between the most important centres of population within the property to facilitate attention to casualties and provision of food and water should be established. Next priority is the clearance and repair of means of access to the outside world; this could be to the project's airstrip, if it has one, or to the nearest government main road. Self-interest dictates that consideration should be given to assisting the government to clear their road from the project to the nearest town if they are unable to cope. After these priorities have been dealt with, a thorough damage survey of the main routes should go on at the same time as harvesting roads are repaired. Priorities may need to be established to rescue nurseries and prevent planting material from being wasted as well as to recover crop before deterioration sets in.

Ensure that road team personnel are aware of their responsibilities in emergencies, that chainsaws can be brought into action quickly and that staff know how to deal with fallen power cables, telephone lines or water pipes safely in the immediate aftermath of the event.

Chapter 6

Rehabilitation of abandoned roads

APPRAISAL OF THE ALIGNMENT

New agricultural or forestry projects are often created upon land which has been logged out first and there will exist on it a number of old abandoned logging roads. The temptation is to use these as the basis of the project's road system. This may or may not be sound common sense.

It is assumed that before starting the project the management will survey the land, either on the ground or by the use of aerial photography, and so can get some idea as to whether the old roads are aligned in such a way as to benefit the project or not. The answer probably is that they are not – after all the loggers will only have been interested in getting their logs out to their sawmill, the nearest main road or export point on a river or seashore – and these are not necessarily on suitable sites for the project's processing centre or the best route out over which the project could dispatch its marketable produce. Furthermore, the logger's roads are likely to have been ill-planned, even for their purpose, and very temporary; they may well have accepted very steep gradients downhill for the loaded log lorries travelling out, knowing that the powerful empty lorries returning to the landings to reload will be little inconvenienced by such adverse gradients. A thorough examination to discover how much, if any, of the old logging system is relevant to the newly proposed developments is therefore recommended. It is suggested that the initial survey of the project's layout is made without any reference to the old roads. When this is completed it can be modified to utilise such of the old roads as are found to be compatible with the new layout and worth rehabilitating. The criteria must be that the modified proposed road system involving rehabilitation of sections of the old logging roads will actually save money compared with the cost of building a completely new layout, and will achieve this without compromising the efficiency of the system. IF THESE CRITERIA ARE NOT MET THEN IT IS BETTER TO IGNORE THE EXISTENCE OF THE OLD ROADS ALTOGETHER except perhaps as a means of temporary access to areas under survey and roads being built. The same strictures must apply when taking over an abandoned estate and its road system.

Such sections of the old road as are acceptable should be checked for gradients to see that they are within the project's limits, that corners and junctions conform, or can be made to conform with the project's criteria and that the placing of bridge and culvert sites is sound.

One advantage that may come with the old roads is that the builders will probably have located a number of quarries that the project can use. These should all be examined.

APPRAISAL OF THE STRUCTURE

Bridges, culverts and drainage

Logger's bridges and culverts are often made of reject species, or the rejected logs of merchantable species, though they may have been forced to use good logs for the beams of the bigger bridges (see Pictures 5.1 to 5.4 on pages 136 & 137). This means that even if the deck beams are sound the abutments are still very suspect. For the loggers this will have been quite acceptable because once the timber has been extracted from their area the roads serve no further commercial purpose for them. They will have no further interest in the future of the land and therefore durability has no merit in their eyes. Very often the bridges and culverts will have collapsed or be in a state of near collapse by the time the land is released for planting and all should be examined before passing over them with a vehicle. Culverts may have just become blocked with the result that the inflowing water is now ponded back. This may have caused softening of the embankments into which the culverts were set with the result that the whole may one day break and wash out during a heavy rainstorm. It is therefore likely that many, if not all, bridges and culverts will have to be replaced and that their replacement will cost rather more than the installation of similar structures on a virgin site unless it is possible to re-align the road to allow the new culverts and bridges to be placed alongside the old ones.

If the abandoned road was not a logger's road but was built of permanent materials it will still be necessary to inspect thoroughly every installation to see how good or bad its condition is and to estimate the repair costs accordingly.

Drains are likely to have become blocked by vegetation, abandoned logs and landslides. This can result in the ponding back of water which will have softened the formation. The first priorities therefore will be to clear the vegetation from the formation, to clean out the drains to the stage that the water can get moving, and drain off any puddles by hand so that rainwater can henceforth get away from the surface and the road can begin to dry out and reconsolidate itself naturally.

Formation

Once the formation has been cleared of weeds and made self draining it will be possible to assess the work that has to be done to upgrade it to a satisfactory standard. Top priority will be to check gradients and corners to see whether re-alignment is necessary, whether hills can economically be cut down to acceptable inclines and corners given acceptable radii for the project's vehicles and equipment to operate on. Next examine for landslides, both those above the road and those eating into the road formation itself. Look for rocks that have not been blown out or dug out but have become prominent above the road surface – estimate the equipment requirements for, and the costs of removing them. Look for soft sections and see whether they are likely to dry out and harden with improvements in drainage or whether the whole section will have to be replaced. Any sections of road driven over swampy ground should be checked for settlement, softening and side slippage and the need for augmenting the fill assessed. From all this a crude estimate of machine time and expenditure required can be made to see whether the rehabilitation is going to be better value for money than making a new road.

IMPLEMENTATION

Much of the work to be done will follow the processes outlined in Chapter 4 and 5 which should be referred to, however:

- It is advisable to use light bulldozers equipped with broad tracks and winches initially because of the risk of becoming bogged down either in softspots resulting from the blockage of drainage or in the work involved in clearing old culvert and bridge sites.
- Select the more imaginative operators for this work which will consist not so much of heavy earthmoving as felling small trees and bush, forming and shaping, creating good roadside and runoff drains – work normally done by the grader but which may be too rough and heavy, or just too sticky for a grader to do when the old road has been freshly screefed of damp undergrowth.
- A track mounted backhoe is likely to prove a most useful machine for digging out the deeper culverts, repairing the bigger drains and clearing landslides.
- The sides of many cuttings and embankments will have become covered by vegetation: this should be preserved and even improved in order to reduce the risks of erosion.

Chapter 7

Staffing and training

EXECUTIVE STAFF

Selection

Apart from the usual characteristics of leadership, responsibility, diligence and honesty essential in any executive, when choosing management for a road programme the candidate should show:

- A marked enthusiasm for the job – for it will involve, on occasions, working long and unsocial hours and bring with it a lot of frustration when the weather is persistently bad.
- Physical fitness for the job, and a willingness to get right to the problem no matter how much walking through the forest or scrambling over difficult ground is involved.
- Creative flair, an ability to conceive shapes, to anticipate the results of both the right way and the wrong way of doing or forming things, and an ability to profit from mistakes made and regard them as useful lessons.
- A feel for machinery and, if not skilled in operating all the machines immediately, at least the willingness to learn and get that skill, to know what they are capable of and to detect quickly when things start to go wrong, well before a major breakdown has occurred.
- A feel for the soil to be worked, with some basic knowledge of geology and the dynamic processes that form a landscape.
- The patience to be a good teacher and the ability to gain respect by example and weld the personnel of the department into a team whose members will be proud to be part of the team and proud of the quality and quantity of the work they produce.
- A well developed sense of initiative, a willingness to tackle work not previously attempted but with the humility to learn in the process, a sharp eye for an opportunity to exploit a situation, take advantage of a spell of good weather or utilise a natural resource, etc.

A good road system manager is as valuable as a good workshop manager and the two are likely to have many characteristics in common. It is an important job: the roads are the estate's arteries and if the transport does not flow freely the estate will soon lose the vigour that characterises an efficient operation.

If the operation is big enough to require middle and junior road team management then many of the qualities listed above will be required from them too. In particular, foremen need to be good leaders and good teachers. If the operation is really big and the road team's machines form the bulk of the workshop's work load, then it may well be desirable that the workshops manager should be responsible to the roads manager until the road development phase is over. This point should be borne in mind when deciding upon the calibre of the road manager to be selected.

Training

Provided that the executive in charge has been well chosen he is likely to benefit from being sent on the occasional generalised management course to keep him abreast with developments in the administrative side of his work. He is likely to ask for technical courses on the machines he has to use, perhaps on the use of explosives and of stone crushing equipment if relevant. This does not mean that he will, for example, handle the explosives himself but he will get up-to-date, know better how the job works, whether the staff delegated to the work are doing it properly, and be better able to act effectively in the event of an emergency. That aside, he is likely to train himself on the job a lot and this should be encouraged.

Much of the training of the other management staff should be effected by the man in charge both on the job and formally by means of lectures, talks or tutorials with the aid of drawings and models. Adequate staff time should be made available for this. Specific skills can often be learnt at courses run by the manufacturers of equipment or suppliers of materials like culverting. Training in the use of explosives is specialised and someone training to be a quarry foreman blaster should attend a formal course with the supplier and pass both the examinations set by the supplier and those set by the local government. Some heavy equipment manufacturers provide courses not only on machine usage but on management, costings and on maintenance. These are usually very worth taking advantage of.

Organisation

The importance of the road construction organisation in the estate management team must be clear right from the start. Without access development cannot proceed. The construction of roads is the sharp edge of development. The appropriate order of operations is therefore that survey should precede roads which should precede logging (if relevant), and preparation and planting. This could mean that initially the project manager is also the road manager since both functions have to commence at the very beginning of a project's life. Indeed this is probably the ideal situation: whoever manages the roads must be intimately involved in the survey and the planning and layout of the whole project. The manager must also be privy to the main development plans so that the steps outlined above proceed in the correct order in each planting block as it is developed. In the short term, problems with the construction of the roads could alter the proposed sequence of planting as much as planting programmes in the long term will dictate the overall extent and shape of the road development programme. These sort of decisions need to be taken at the very highest level ... ON SITE.

The road management will need the power and status to be able to enforce discipline in the use of the roads. Careless use of the roads during construction or repair can result in costly damage out of all proportion to any advantage to the project overall occasioned by such usage. To maintain output the workshops staff must work very closely with the roads management to ensure that servicing of road building machinery is done in such a way as to minimise the loss of productive machine time.

It is clear therefore that, in the early stages of the development of a project, the project manager, if not himself managing the road work, must delegate to a person with whom he can work very closely indeed, and to whom he is prepared to give all the authority, support and information necessary to ensure a satisfactory result. Once the essential framework of the communications system is built, substantial alterations to the basic fabric are going to be very expensive to make.

When the road construction phase (which may only last for a few years) has finished, the needs for road management will change. Instead of being a powerful development unit in its own right the road team will become primarily a maintenance unit. As such it will be better merged into the plantation management, probably at divisional level. On large projects there may also be a need for one independent centrally controlled team coping with main roads and emergencies, part of the same group that (like the head office, workshops, buildings upkeep and similar management divisions), services the main productive unit, i.e. the plantings plus processing centre. In recruiting and organising road management it is necessary to anticipate these changes to avoid redundancy and to make the optimum use of the skills learned and historical knowledge gathered during the early phases of the programme. Consider, therefore, putting plantation management cadets into road team management both to get enthusiastic youngsters involved – so ensuring that they have the necessary knowledge later to be able to maintain their own plantation's roads – and, should the need arise in the future, to ensure they will be capable of starting new projects. During the construction phase, from amongst the team foremen, look out for the really good steady man with great pride in his work who could head the main road maintenance team in due course.

A road team manager having a substantial programme to carry out requiring two or three construction teams, anything from say six to fifteen bulldozers all told, will need a survey team foreman, a foreman for each road team and a good clerk to keep the records. The foremen will each need pickups, preferably with four-wheel-drive, and the manager will also need a strong four-wheel-drive vehicle. The pickups need to be able to carry men, tools and spares such as wheels for loaders and graders and track chains for bulldozers. Inadequate vehicles will soon get broken. The clerk will, at the least, need a good motorcycle to enable him to get around and check recorders, gravel deliveries and the like. A bridge construction team foreman may be required. If a large crusher is operated, a quarry foreman will be necessary who may also need to be a qualified blaster. A quarry clerk to record loads of stone and keep explosives records may be required. The clerks have to be willing to go out into the forest in the course of their work. They should also cope with the wages and work allocations of the labour and costs and job allocations for their machines. Good men should be recruited as foremen with a view to either putting them in charge of road maintenance or transferring them into plantation or other work later, where their experiences, and the attitudes to work that they will hopefully have gained in the road construction teams, will be of considerable value.

A large project may make the use of mobile housing advisable for the construction teams, both to save on travelling time and to provide the machinery with some sort of security in the form of the presence of the operators nearby at night. If this is done the foremen should also live with their teams. On a small project this may not be necessary. Where operators are taken out each morning and returned each night the unfailing provision of transport on time, both in the mornings and in the evenings, is vitally important for morale. Nothing upsets operators more than to find that with night having fallen after a long hard day, or with the rain belting down so that work is no longer possible, they are faced with a walk home of several miles. It is management's responsibility to see that this does not happen, loyalty must be a two-way thing if it is to have any real meaning.

OPERATORS

Selection

If there is time, and if there are one or two foremen who are good at teaching, it is better to recruit keen youngsters and train them as required for the team's needs. This avoids unpleasant experiences with operators who have picked up bad habits elsewhere. If that route is not open, then the selection of the first few operators, who are likely to be much involved in training others, becomes critical.

Whether selecting untrained or experienced operators remember that handling heavy machinery requires endurance, so physical fitness is imperative. A responsible attitude is paramount; irresponsibility leading to breakdowns or wasted work can be very costly indeed. Again the ability to conceive shapes and the effects of road form on drainage and traffic movement is valuable. Men who are skilled at making other things, whether it be in wood, metal or stone, are more likely to have this attribute. For trainees, prior experience with other machinery is not necessary (one of the best bulldozer operators ever to work with the writer came from the interior of Borneo and had never seen any sort of machine before in his life. He was very good at carving wood, boat building and making utensils out of materials from the forest and asked for the chance to learn; within three months he was already one of the top two all round operators on the project). Look for the quiet character, not the show-off who merely wants to revel in the power, size and potential for destruction of his machine.

Training

Some suppliers of heavy equipment, compressed air drilling equipment, crushers and such like run courses for operators. They tend usefully to concentrate on "do's", "do not's" and give the operator some idea of, and respect for the complexity of his charge, and its value. It is impossible in a few days to inculcate operating skills. This takes weeks. At an early stage, and if readily available, these courses are worth having, particularly the more specialist ones. There are also useful courses for the conversion of operators from one machine type to another.

Most training is done on the job. Many operators start to learn either as hookmen or second drivers to other operators who are already skilled and, if the two get on well together, this can be very satisfactory. Special training needs to be given by the road team foreman, both formally and on the job, to cover daily checks and procedures (refuelling, cleaning down, inspections for fractures, breakages etc, starting-up and shutting-off the machines, closing off work at night or when rain interferes with work). As and where opportunity arises, he should demonstrate special techniques of operating (for example, the use of a winch or the super-elevation of a corner). A limited spell in the workshops under the sharp eye of the chief heavy equipment mechanic can broaden the operator's mind and make him more aware of the strengths and weaknesses of his machine.

It will pay to train operators to use more than one type of machine because:

- This will give the operator a better idea of the capabilities of machines different from his own and therefore an appreciation of how the different machines can better complement one another in a team.
- Operators often show unexpected flair on one or other machine type, this will give management a chance to recognise and place outstanding ability to the company's best advantage.
- It is very useful in emergency to have operators who can handle a wide range of machines.

Organisation

In most tropical countries the cost of the operator of a machine will be infinitesimal compared with all the other costs involved (depreciation, loss of interest on capital, spares, repairs, fuel and lubricants). As has been mentioned in Chapter 3, there is therefore considerable potential for improving output and cutting unit costs by putting a lead and second operator on each of the more demanding machines, graders and bulldozers particularly. The machines can then be operated for all the daylight hours there are with the operators both being paid a full day plus overtime and working turn and turn about every two hours or so. There is then no excuse for the machine stopping for lunch or any other breaks, or for the operators losing concentration because they are hot and tired. It also facilitates training and substitution for leave or illness. It must be quite clearly understood that the lead operator is responsible for the machine and also for the second operator. The pair of them have to be chosen with care to be sure they get on well together. It pays to let the lead operator choose his own second operator: this underlines for him his responsibility and stops any attempt to try and shed blame for something that goes wrong.

In a construction team, consisting of from say two to five bulldozers, one operator should become the team lead operator. He should be the most skilled, respected and responsible man of the group regardless of the size of the machine he is operating, and be capable of deputising for the team foreman for short periods. He is the sort of person that one would have an eye on for possible promotion. Remember that promotion "in house" is a great morale builder if the right sort of men are selected.

Teams should be worked as teams as far as is practical – this is not to say that a machine should never be detached for short assignments separately – but because being a group of complementary tools for a job they will work better together and because the team spirit will always be at its most effective when the team is together. If this is not the case then find out why, quickly, and repair the situation. A team which is not a team, because the members of it do not get on together, will soon waste a great deal of money for the project.

Chapter 8

Accounting and cost control

GENERAL

It is not intended here to give detailed costing and accounting systems – all organisations will have their own methods of doing these things and the road building and maintenance accounts will have to conform. It is only intended here to look at ways of getting the basic information out so that it can be utilised by the project's systems.

Because of the differences in the way companies treat their accounts, also because of differences in the physical conditions that form the environment in which the road construction teams operate, comparisons between companies or projects of cost per unit of road constructed are frequently valueless. No attempt will be given here to indicate what a kilometre of road should cost. This does not invalidate the concept of gathering costs per unit as a project proceeds and using this, with some application of commonsense, for the purposes of making project estimates for cash flow purposes. However, estimates should be regarded as estimates: intelligent guesswork that may be way out if the unexpected happens.

RECORDING MACHINE TIME FOR COSTING

Meters and clocks

The meters provided on many items of heavy equipment are intended for indicating when routine service and maintenance is due. They are not accurate enough for use as a means of recording actual time worked. For this a proper time recorder is needed. A simple recorder capable of recording a full twenty-four hours on a waxed card can be fitted and should give several years service.

In use the recorder, which has a stylus in contact with a circular disc that revolves once in a day, will show a thin line when the machine is not working at all, a thin slightly wobbly line when the engine is idling and the machine is static, and a thick, well scored line when the machine is working. The card is marked in hours and minutes and therefore it is possible to see exactly when the machine was not in use, when idling and when working. Time can be calculated accordingly. The information on the clock discs will need to be augmented by notes from the operator and/or foreman to show what job it was on and when the machine transferred from one job to another. This information will then enable three types of data to be compiled:

- A record of the time worked by the operators, including their overtime, which can be used to support paysheets.
- A record of total gross time worked as a basis for working out the costs of operating the machine per hour.
- A record of total nett time employed on each separate job tackled as a basis for the allocation of charges for the use of the machine by account heads.

Implied above is the suggestion that the thorny question of what to do with transit time between jobs is solved by disregarding it for the purposes of charging out to account heads though it must be retained for the purposes of working out pay and the running costs of the machine. Thus the basis for the rates for charging purposes will be set by dividing the total operating costs by the total nett hours worked and multiplying them by the number of hours allocated to each individual job. Total nett time chargeable will always be less than total gross time worked, since gross time worked will include time driving the machine from one worksite to another and any other manoeuvres performed that are not directly productive. Records of total time worked will still be required as they indicate, supplemented with notes on downtime due to bad weather, the availability of the machine and therefore its reliability. However, if the machine is being hired from a contractor on an hourly rate he can fairly expect to charge for transit time between jobs. The contractor may also levy a positioning charge to cover the costs of bringing his machines to the project and taking them away afterwards. This does not prevent the project from recording nett productive time and using this as a basis for apportioning the contractor's charges between account heads.

To avoid problems arising from constantly varying charge rates, and some account heads being unduly penalised because, by chance, the work on them was done when an exceptional cost was incurred (for example a bulldozer track change), estimated charge rates (often referred to as standardised rates) can be applied for a quarter or even a half year. These can be adjusted up or down, after actual costs are known, to compensate for variations with the objective of being neither in credit nor debit to any significant amount at the end of the financial year. The principle of standardisation of rates is much easier to operate if the charges can be averaged over several identical machines.

Individual costs per hour based on gross time worked should be recorded and kept with the machine's historical records as a means, along with availability records, of finding out when the machine has eventually become uneconomic to run.

Whereas the costs of dumpers are usually kept and charged for in hours, the costs of tippers and lorries are more commonly charged on the basis of distance run. This is not necessarily a good idea – since this implies that time costs nothing there is no penalty for holding onto a lorry for trivial reasons and so project staff have no incentive to get a lorry away promptly for other work. Where estate transport, including road team tippers, is used for hauling labour and supplies to the fields before settling down to the main work of the day, there is much to be said for fitting recorders to the tippers and charging by the hour. The records will show where the lorries have been wasting their time and the costs will show up in excessive transport cost allocations against the work of the staff responsible. Senior management can then take remedial action.

Compiling the records

The recorder charts should be collected and entered by the foremen or their clerks every day into the periodic (weekly, fortnightly or monthly according to the project's way of operating) records. The labour side of the records should show attendance and overtime worked and, if relevant, any output figures concerning bonuses or pieceworks. The machine side of the records will show total gross time worked split into chargeable and non-chargeable time. Separate notes should be kept on downtime showing whether this was due to absence of operators, to weather, to breakdown or servicing. It is assumed for these purposes that public holidays or other days when the project does not work are deducted from the potential total of working hours before any comparisons are made.

At the end of the period the labour records are used for pay preparation. The operator's total time worked should tally with his machine's total time worked plus any allowance given to the operator for time spent on refuelling, inspecting and cleaning down the machine, travelling time, etc. The machine time should show chargeable time by account heads and total gross operating time. Meter time should be recorded as a check: although the figure will not be the same as true time any large and unusual discrepancy should attract management's attention. It also serves as a warning in advance of when the next major service is due so that arrangements can be made that minimise downtime.

Records of fuel, lubricants and hydraulic oil issued should be kept by the road team clerk for each individual machine and these consolidated along with spares, repairs and maintenance costs, onto the machine's historical record. These should be shown both as totals and as costs per hour for the period and to date. Spares, repairs, hydraulic oil and lubricants costs per hour can be expected to fluctuate considerably in the current period column but fuel costs should not. Lubricant costs should not change greatly over time in the to date costs but spares and repairs costs can be expected to start off at a very low level, rising with time until a major overhaul has been completed. They should then level off or fall somewhat but will rise to another peak at the time of the next overhaul. When to date costs for spares and repairs rise to be consistently around or above depreciation levels, the machine is no longer economic to run.

Each machine's historical record should commence from the day it starts service and go on until it is disposed of. Not only should it have the costs summarised onto it month by month but written notes should be added to cover all major services, major repairs and any accidents befalling it. Histories should be grouped by machine types and, where there are several units of one type, identical failures that occur on more than one machine can usefully be drawn to the manufacturer's attention with either the aim of getting compensation or the modification at manufacturer's expense of all machines of that type. This, hopefully, will avoid repetition of the failure and the inevitable downtime that goes with it.

The depreciation question

The depreciation rate is based on the anticipated life of the machine. It can be expressed as the cost of purchase divided by the economic life expectancy, five years or 10,000 hours being fairly typical figures for a bulldozer. Whether to depreciate in hours or years will depend upon the project's accounting policy. If depreciation is to be included into the charges it is better to base it on an hourly rate. If the depreciation is taken into the accounts as an overhead charge then a yearly rate is likely to be preferred. It is prudent to consider the conditions under which the machine will work before deciding just how many thousand hours service it can be expected to give, and then to set the depreciation period. No less relevant is the quality of maintenance that can be given. Generally, amongst heavy earthmoving equipment, tracked machines wear out more quickly than wheeled machines. Typical figures are as follows:

- Bulldozers   10,000 hours
- Tracked loaders, tracked backhoes   12,000 hours
- Wheeled loaders, wheeled backhoes, steel rollers   15,000 hours
- Graders, articulated dumpers, rubber rollers   20,000 hours

If conditions are particularly harsh the figures should be downgraded by about twenty percent. The number of hours that can be clocked up in a year will depend both upon the amount of work required of the machine and the limitations upon one's ability to achieve that requirement imposed by the weather. These factors should be estimated when the depreciation rates and periods are decided upon.

COSTING ROADS

Construction costs

There is considerable point in recording the costs per mile or kilometre of some of the roads as they are completed on a large project. Concentrate on what might be called standard roads, for example harvesting roads where, over several years, roads to the same specification will be built on similar ground for successive plantings. The information will enable future estimates to be made more accurately and may indeed be able to indicate where changes of method or equipment do, or do not, cut costs.

The costings must be comparable: by that it is meant that there must be no change in policy affecting them over the period. For example, there must be a decision taken (purely for comparison and estimating purposes) as to whether the costs of felling bush to make a harvesting road are to be regarded as a land preparation charge or a road charge and, once the decision has been taken it must not be changed.

A costing can only really be effected upon the completion of a sector of road and on the basis that, in the work allocations for both labour and machines, it has been kept as a separate account head. It is futile to try to attempt to give road costs per unit distance while the sector is still half completed. Once finished, the total cost can be divided by the distance to give cost per kilometre or mile. It is reasonable to break down the sector cost into major stages, for example the cost of the formation completed by the bulldozers including all drainage and culverts, preliminary compaction and grading; the cost of surfacing, grading and final compaction.

The costing of unique sectors of road, unlikely to be repeated again, is not likely to be of much value except as a justification for building an "expensive" short route instead of a "cheaper" (both terms used in respect of costs per unit length) but longer route. However, on the rule of thumb basis that the annual cost of upkeep of a road is likely to be about ten per cent of the cost of construction (but maybe twice that in the first year), it is useful in preparing upkeep estimates for that sector until actual experience enables the estimates to be refined.

If road costs are to be tied into field costs, the roads being regarded as an essential part of a productive planting, then maps will have to be made showing which roads belong to which block; not necessarily so simple as it may seem because the roads will often form block boundaries for convenience and one will probably wish to avoid the trouble of charging half a road's costs to one block and half to the other. The allocation of roads to blocks must also be made with some thought for practicality in the field: it is pointless to expect a grader driver to stop halfway down a road because a field boundary happens to cross it at that point. He must be able to work sectors logically, between one junction and another for example, and have a clear cut off point at which he is able to put time worked down to the one field or the other in his notebook.

Maintenance costs

Given average weather, the costs of maintenance are likely to be up to twice as high in the first year after construction as in subsequent years. This is because the soil has been freshly disturbed during construction and any anti-erosion plantings effected will have had no time to become established and fully effective. Erosion is therefore likely to be more severe than it will be later when things have settled down. When recording actual costs of upkeep, sector by sector for main roads, block by block for harvesting roads, this should be borne in mind so that only second and subsequent year maintenance is regarded as being a sound indicator for the future.

It should be realised that variations in weather from one year to another can have very significant effects on both road construction and maintenance costs. It is essential that estimates be regarded not as a limitation but a prudent provision; adequate standards must be maintained. If this results in underspending, all well and good, but if it is not going to be enough in a wet year, then adequate additional money must be made available. Alternatively, in the case of construction work, the programme can perhaps be curtailed until there is an improvement in the weather (thus underlining the need to keep the road development work well ahead of plantation development as a matter of principle). This indicates that savings in expenditure one year are not necessarily sound reasons for reducing the estimate for the next year. Equally, heavy expenditure on maintenance in a bad year is not justification for expecting expenditure at the same high level in the following year. It has to be remembered that it is going to cost a great deal more to put right the results of neglect in a bad year when it is tackled a year later than it would have done had the maintenance been done properly at the time. Such neglect will probably have resulted in damage to drainage and the development of gulleys, potholes and softspots. Apart from the damage to the roads, it will have done no good to the vehicles using them either. So the maintenance of the correct standards may result in sharply fluctuating costs between one year and another but it is good cost control to allow them to do so and thereby prevent the project from incurring unnecessary costs arising from deliberate neglect, no matter how well intentioned. What is needed is that management should, by being on site and supervising operations, know whether they are getting value for money or not.

Further reading and useful addresses

FURTHER READING

Rural road building and upkeep

Building Roads by Hand. By J. Antoniov, P. Guthrie and J. de Veen.

Low Cost Roads in the Rural Environment.
Papers presented at a conference on rural road building held in Birmingham in April, 1994. Available from the Forestry Engineering Group of the Institution of Agricultural Engineers through:

G. Friedman, Forest Enterprise
231 Corstorphine Road
Edinburgh EH12 7AT
Scotland, UK
Tel.: IDD+44 (0)131-334-0498

Bridges

Pre-Fabricated Modular Wooden Bridges, Part 1, General Description.
This little booklet describes briefly the form and construction of modular wooden bridges of 9 to 24 metres span capable of carrying lorries of up to 32 tons weight; does not provide enough information for construction to be undertaken but does enable an assessment to be made of the suitability of this form of bridge for any given project (Parts II, III and IV are for restricted circulation only). Available from:

Timber Research and Development Association (TRADA)
Hughenden Valley, High Wycombe
Buckinghamshire, HP14 4ND, UK. (Telex 83292)

Timbers of the World, Volume I: Timbers of Africa, South America, India, and the Far East; Volume II: Timbers of Japan, Australasia, Western Europe, North America and Central America. Also available in the form of " Red Booklets" of which there are ten, each covering a particular region. These give the properties of every usable timber in the areas covered. Available from TRADA, see address above.

Piers, Abutments and Formwork for Bridges. By J.R. Robinson, available in French or English. Whilst primarily written for the serious civil engineer building bridges, there are a lot of references to mold bridges in France and the reasons for their failure or continued existence from which much can be learned. Available from:

Crosby Lockwood & Son Ltd.
26, Old Brompton Road
London SW7, UK.

Quarrying and Rockbreaking. Compiled by D. Lester.

Road maintenance

Road Maintenance Handbook, Volume I: Maintenance of roadside areas, drainage structures and traffic control devices and *Volume II: Maintenance of unpaved roads.* A compact pair of little books full of diagrams and drawings specifically for African conditions of unsealed public road upkeep. Available in English or French from:

Transport and Road Research Laboratory
Department of Transport
Old Wokingham Road, Crowthorne
Berkshire, GR11 6AU, UK.
Tel.: IDD+44 (0)1344-77-3131

Soil stabilisation

Soil Stabilisation with Cement and Lime. A state of the art review by P.T. Sherwood of the Transport Research Laboratory, Department of Transport, UK. This is a comprehensive examination of various methods of soil stabilisation worldwide. It has also a list of over 100 references to other works on various aspects of the subject. Available from:

H.M. Stationery Office
St Crispin's House
Duke Street
Norwich NR3 1PD, UK.
Tel.: IDD+44 (0)171-873-9090

Survey

Estate Surveying by R.B. Perkins, a practical textbook in basic surveying written for trainee estate managers. Available from:

> The Incorporated Society of Planters
> P.O. Box 10262, Kuala Lumpur, Malaysia
> Tel.: IDD+60 (0)3-242-5661

Sawmilling

Saws and Sawmills for Planters and Growers by John M. Morris (tel.: 040-382-2676). Covers the selection, use and upkeep of handsaws, chainsaws, mobile and transportable sawmills useful in project development situations. Saws, suitable both for forest clearance and the preparation of lumber for bridges and culverts etc. for roadwork, are covered, amongst other applications. Available from:

> Avebury, Ashgate Publishing Ltd.
> Gower House, Croft Road, Aldershot
> Hampshire GU11 3HR, England, UK.
> Tel.: IDD+44 (0)1252-331551

USEFUL ADDRESSES

Engineering and Mechanics Innovation and Research (EMIR) Ltd
'Baghdad', Dee Road, Talacre, Clwyd CH8 9RS, UK.
(See Polewood Bridges section, page 174)

The Institution of Agricultural Engineers
(Forestry Engineering Group)
231 Corstorphine Road,
Edinburgh EH12 7AT, Scotland, UK.

Orr's Ghat road tracer is available from:
E.J. Motiwalla
36 Persiaran Ampang
55000 Kuala Lumpur, Malaysia

Timber Research & Development Association
Hughenden Valley
High Wycombe, Buckinghamshire, HP14 4ND, UK.

Transport and Road Research Laboratory
Department of Transport, Old Wokingham Road
Crowthorne, Berkshire, GR11 6AU, UK.

Index